AF596532

Olfaction

Olfaction: From Genes to Behavior

Editor

Edgar Soria-Gómez

MDPI • Basel • Beijing • Wuhan • Barcelona • Belgrade • Manchester • Tokyo • Cluj • Tianjin

Editor
Edgar Soria-Gómez
University of the Basque Country UPV/EHU
Spain

Editorial Office
MDPI
St. Alban-Anlage 66
4052 Basel, Switzerland

This is a reprint of articles from the Special Issue published online in the open access journal *Genes* (ISSN 2073-4425) (available at: https://www.mdpi.com/journal/genes/special_issues/olf_gen).

For citation purposes, cite each article independently as indicated on the article page online and as indicated below:

LastName, A.A.; LastName, B.B.; LastName, C.C. Article Title. *Journal Name* **Year**, *Article Number*, Page Range.

ISBN 978-3-03936-621-7 (Hbk)
ISBN 978-3-03936-622-4 (PDF)

Contents

About the Editor

Edgar Soria-Gómez has been involved in neuroscience and endocannabinoid research since 2001, when he joined Dr. Oscar Prospéro-García's laboratory, the leading Mexican lab devoted to the study of the endocannabinoid system. There, he completed his bachelor and doctoral studies (2002–2009). He obtained his Ph.D. in Biomedical Science in 2009 from the Universidad Nacional Autónoma de México (UNAM, Mexico). During that period, in 2007, he had the opportunity to spend several months abroad in Liverpool, UK, in Professor Tim C. Kirkham's laboratory, the first to demonstrate the appetite-stimulating properties of endocannabinoids. Then, in 2009, after being awarded the Fyssen Foundation Fellowship, he started his postdoctoral research as part of Giovanni Marsicano's team, at the NeuroCentre Magendie (Bordeaux, France), one of the most prestigious international environments for the study of endocannabinoid functions. In 2018, he obtained the Ikerbasque Research Fellowship at the Achucarro Basque Center for Neuroscience at the University of the Basque Country in Bizkaia, Spain. There, he is developing his line of research, which mainly focuses on the study of the endocannabinoid system and the identification of the brain circuits which control physiological and pathological states.

Editorial

Special Issue "Olfaction: From Genes to Behavior"

Edgar Soria-Gómez [1,2,3]

1 Department of Neurosciences, University of the Basque Country UPV/EHU, 48940 Leioa, Spain; edgarjesus.soria@ehu.eus or edgar.soria@achucarro.org
2 Achucarro Basque Center for Neuroscience, Science Park of the UPV/EHU, 48940 Leioa, Spain
3 IKERBASQUE, Basque Foundation for Science, 48013 Bilbao, Spain

Received: 12 June 2020; Accepted: 15 June 2020; Published: 15 June 2020

The senses dictate how the brain represents the environment, and this representation is the basis of how we act in the world. Among the five senses, olfaction is maybe the most mysterious and underestimated one, probably because a large part of the olfactory information is processed at the unconscious level in humans [1–4]. However, it is undeniable the influence of olfaction in the control of behavior and cognitive processes. Indeed, many studies demonstrate a tight relationship between olfactory perception and behavior [5]. For example, olfactory cues are determinant for partner selection [6,7], parental care [8,9], and feeding behavior [10–13], and the sense of smell can even contribute to emotional responses, cognition and mood regulation [14,15]. Accordingly, it has been shown that a malfunctioning of the olfactory system could be causally associated with the occurrence of important diseases, such as neuropsychiatric depression or feeding-related disorders [16,17]. Thus, a clear identification of the biological mechanisms involved in olfaction is key in the understanding of animal behavior in physiological and pathological conditions.

The olfactory system is a one-in-a-kind sensory system, because olfactory sensory neuro-epithelial neurons located in the nasal cavity and expressing specific odor receptor send direct projections to the main olfactory bulb (MOB), without a thalamic relay. Within the MOB, the processing of olfactory information and their relay to higher brain regions is guaranteed via a vast heterogeneity of cell-types. The work of Sanchez-Gonzalez et al. [18] defined the distribution and the phenotypic diversity of olfactory bulb interneurons from specific progenitor cells, focusing on their spatial origin, heterogeneity, and genetic profile. Fengyi Liang [19] contributes to the study of the cytoarchitecture of olfactory circuits, by reviewing the relevance of the cellular link between the olfactory receptor neurons (ORN) and the olfactory sustentacular cells (OSC). Indeed, the different olfactory functions could rely on complex cellular interactions [20], which are also regulated by neuromodulatory systems. Among them, the endocannabinoid system is emerging as a link between olfactory information and behavioral processes (e.g., memory and food intake), as reviewed here by Terral et al. [21]. Olfactory structures are the target of peripheral signals sensing the nutritional status of the organism [22], consequently affecting feeding behavior. Wu et al. [13] describe how the mitral cell (MC) activity in the MOB changes when there is a negative energy balance. Interestingly, such changes are related to impairment in olfactory discrimination. Thus, olfactory circuits represent a very interesting model system to understand general rules of information processing in the brain necessary for the species survival. In this context, several studies show that olfactory cues could also be determinant for partner selection and sexually driven behavior [2,23,24]. The work of Fraichard et al. [25] shows that the odorant-degrading enzymes (ODE) participate in mate selection. In particular, they demonstrate that the UDP-glycosyltransferase (UGT36E1) expressed in the olfactory sensory neurons (OSN) of the *Drosophila* is involved in sex pheromone discrimination. Furthermore, Liu et al. [26] present a complete review of the genetics and evolution of chemosensory detection, highlighting its potential role in modulating physiological processes, including pheromone detection. As the authors mention, chemosensitivity represents a key function in a primary common universal mechanism of eukaryote and prokaryote cells and in their interactions with the changing environment.

Interestingly, sensing of chemical signals, in particular olfactory cues, could have a global influence at many different levels, from basic survival mechanisms to economic impacts in modern society. For example, the parasitoid wasp Ashmead, *Diachasmimorpha longicaudata* is used as a control agent in pest management to suppress fruit flies. Here, Tang et al. [27] performed a detailed transcriptome analysis showing that olfactory genes of the parasitoid wasps are expressed in response to their hosts with different scents. By using a similar methodological approach, Wang et al. [28] contribute to answering an open question about whether males and females possess the same abilities to sense odorants. Several studies have suggested that external stimuli, including courtship songs, colors and chemosensory cues, could be determinant for sex-specific behaviors. The authors reveal that, in zebrafish, chemosensory receptor genes are more expressed in males than in females, suggesting the existence of sex-specific neuronal circuits. In this sense, Tasmin L. Rymer [9] reviews the existing literature about the influence of olfactory cues in rodent paternal behavior, highlighting the role of ten genes mainly involved in aggressive responses towards intruders and pups recognition. In summary, this Special Issue reflects the state-of-the-art in olfactory research, opening new possibilities for interdisciplinary studies, from genes to behavior.

Funding: This research received no external funding.

Conflicts of Interest: The author declares no conflict of interest.

References

1. Stevenson, R. Phenomenal and access consciousness in olfaction. *Conscious. Cogn.* **2009**, *18*, 1004–1017. [CrossRef] [PubMed]
2. Kringelbach, M.L. *The Pleasure Center: Trust Your Animal Instincts*; Oxford University Press: New York, NY, USA, 2009; p. 291.
3. Hoover, K. Smell with inspiration: The evolutionary significance of olfaction. *Am. J. Phys. Anthropol.* **2010**, *143* (Suppl. 51), 63–74. [CrossRef] [PubMed]
4. Trellakis, S.; Fischer, C.; Rydleuskaya, A.; Tagay, S.; Bruderek, K.; Greve, J.; Lang, S.; Brandau, S. Subconscious olfactory influences of stimulant and relaxant odors on immune function. *Eur. Arch. Oto-Rhino-Laryngology.* **2012**, *269*, 1909–1916. [CrossRef]
5. Doty, R.L. Odor-guided behavior in mammals. *Experientia* **1986**, *42*, 257–271. [CrossRef]
6. Johansson, B.R.; Jones, T.S. The role of chemical communication in mate choice. *Biol. Rev. Camb. Philos. Soc.* **2007**, *82*, 265–289. [CrossRef]
7. Fletcher, N.; Storey, E.J.; Johnson, M.; Reish, D.J.; Hardege, J.D. Experience matters: Females use smell to select experienced males for paternal care. *PLoS ONE* **2009**, *4*, e7672. [CrossRef]
8. Dias, B.; Ressler, K.J. Parental olfactory experience influences behavior and neural structure in subsequent generations. *Nat. Neurosci.* **2014**, *17*, 89–96. [CrossRef]
9. Rymer, T.L. The Role of olfactory genes in the expression of rodent paternal care behavior. *Genes* **2020**, *11*, 292. [CrossRef]
10. Rolls, E.T. Taste, olfactory, and food texture processing in the brain, and the control of food intake. *Physiol. Behav.* **2005**, *85*, 45–56. [CrossRef] [PubMed]
11. Stafford, L.; Welbeck, K. High hunger state increases olfactory sensitivity to neutral but not food odors. *Chem. Sens.* **2011**, *36*, 189–198. [CrossRef] [PubMed]
12. Aimé, P.; Duchamp-Viret, P.; Chaput, M.A.; Savigner, A.; Mahfouz, M.; Julliard, A.K. Fasting increases and satiation decreases olfactory detection for a neutral odor in rats. *Behav. Brain Res.* **2007**, *179*, 258–264. [CrossRef] [PubMed]
13. Wu, J.; Liu, P.; Chen, F.; Ge, L.; Lu, Y.; Li, A. Excitability of neural activity is enhanced, but neural discrimination of odors is slightly decreased, in the olfactory bulb of fasted mice. *Genes* **2020**, *11*, 433. [CrossRef] [PubMed]
14. Krusemark, E.A.; Novak, L.R.; Gitelman, D.R.; Li, W. When the sense of smell meets emotion: Anxiety-state-dependent olfactory processing and neural circuitry adaptation. *J. Neurosci. Off. J. Soc. Neurosci.* **2013**, *33*, 15324–15332. [CrossRef] [PubMed]
15. Buron, E.; Bulbena, A. Olfaction in affective and anxiety disorders: A review of the literature. *Psychopathology* **2013**, *46*, 63–74. [CrossRef]

16. Rapps, N.; Giel, K.E.; Söhngen, E.; Salini, A.; Enck, P.; Bischoff, S.C.; Zipfel, S. Olfactory deficits in patients with anorexia nervosa. *Eur. Eat. Disord. Rev. J. Eat. Disord. Assoc.* **2010**, *18*, 385–389. [CrossRef]
17. Oral, E.; Aydin, M.D.; Aydin, N.; Ozcan, H.; Hacimuftuoglu, A.; Sipal, S.; Demirci, E. How olfaction disorders can cause depression? The role of habenular degeneration. *Neuroscience* **2013**, *240*, 63–69. [CrossRef]
18. Sánchez-González, R.; Figueres-Oñate, M.; Ojalvo-Sanz, A.C.; López-Mascaraque, L. Cell progeny in the olfactory bulb after targeting specific progenitors with different ubc-startrack approaches. *Genes* **2020**, *11*, 305. [CrossRef]
19. Liang, F. Sustentacular cell enwrapment of olfactory receptor neuronal dendrites: An update. *Genes* **2020**, *11*, 493. [CrossRef]
20. Yamaguchi, M. Functional sub-circuits of the olfactory system viewed from the olfactory bulb and the olfactory tubercle. *Front Neuroanat* **2017**, *11*, 33. [CrossRef]
21. Terral, G.; Marsicano, G.; Grandes, P.; Soria-Gómez, E. Cannabinoid control of olfactory processes: The where matters. *Genes* **2020**, *11*, 431. [CrossRef]
22. Julliard, A.; Al Koborssy, D.; Fadool, D.A.; Palouzier-Paulignan, B. Nutrient sensing: Another chemosensitivity of the olfactory system. *Front. Physiol.* **2017**, *8*, 468. [CrossRef] [PubMed]
23. White, T.L.; Cunningham, C. Sexual preference and the self-reported role of olfaction in mate selection. *Chemosens. Percept.* **2017**, *10*, 31–41. [CrossRef]
24. Kromer, J.; Hummel, T.; Pietrowski, D.; Giani, A.S.; Sauter, J.; Ehninger, G.; Schmidt, A.H.; Croy, I. Influence of HLA on human partnership and sexual satisfaction. *Sci. Rep.* **2016**, *6*, 32550. [CrossRef] [PubMed]
25. Fraichard, S.; Legendre, A.; Lucas, P.; Chauvel, I.; Faure, P.; Neiers, F.; Artur, Y.; Briand, L.; Ferveur, J.F.; Heydel, J.M. Modulation of sex pheromone discrimination by A UDP-Glycosyltransferase in *Drosophila melanogaster*. *Genes* **2020**, *11*, 237. [CrossRef]
26. Liu, G.; Xuan, N.; Rajashekar, B.; Arnaud, P.; Offmann, B.; Picimbon, J.F. Comprehensive history of CSP genes: Evolution, phylogenetic distribution and functions. *Genes* **2020**, *11*, 413. [CrossRef]
27. Tang, L.; Liu, J.; Liu, L.; Yu, Y.; Zhao, H.; Lu, W. De novo transcriptome identifies olfactory genes in *Diachasmimorpha longicaudata* (Ashmead). *Genes* **2020**, *11*, 144. [CrossRef]
28. Wang, Y.; Jiang, H.; Yang, L. Transcriptome analysis of zebrafish olfactory epithelium reveal sexual differences in odorant detection. *Genes* **2020**, *11*, 592. [CrossRef]

Article

Cell Progeny in the Olfactory Bulb after Targeting Specific Progenitors with Different UbC-StarTrack Approaches

Rebeca Sánchez-González, María Figueres-Oñate †, Ana Cristina Ojalvo-Sanz and Laura López-Mascaraque *

Department of Molecular, Cellular and Development Neurobiology, Instituto Cajal-CSIC, 28002 Madrid, Spain; rebeca@cajal.csic.es (R.S.-G.); maria.figueres@gen.mpg.de (M.F.-O.); anacris23@cajal.csic.es (A.C.O.-S.)

* Correspondence: mascaraque@cajal.csic.es

† Present address: Max Planck Research Unit for Neurogenetics, Frankfurt am Main, 60438 Frankfurt, Germany.

Received: 30 January 2020; Accepted: 11 March 2020; Published: 13 March 2020

Abstract: The large phenotypic variation in the olfactory bulb may be related to heterogeneity in the progenitor cells. Accordingly, the progeny of subventricular zone (SVZ) progenitor cells that are destined for the olfactory bulb is of particular interest, specifically as there are many facets of these progenitors and their molecular profiles remain unknown. Using modified StarTrack genetic tracing strategies, specific SVZ progenitor cells were targeted in E12 mice embryos, and the cell fate of these neural progenitors was determined in the adult olfactory bulb. This study defined the distribution and the phenotypic diversity of olfactory bulb interneurons from specific SVZ-progenitor cells, focusing on their spatial pallial origin, heterogeneity, and genetic profile.

Keywords: Olfactory interneuron; heterogeneity; StarTrack; development; neural progenitor cell; cell fate; in utero electroporation

1. Introduction

The mammalian olfactory system is composed of the olfactory epithelium (OE), olfactory bulb (OB), and olfactory cortex (OC). The rodent OB is organized into six layers that contain distinct cell populations and that are essentially made up of two types of neurons, interneurons, and projection/output neurons [1,2]. Using the Golgi method, Santiago Ramón y Cajal described the layers in the OB and its components more than a century ago (Figure 1A reproduces an original drawing of Cajal [3]; reviewed in [4]). His morphological studies on the OB provided the basis to define the neurons present in this structure and when UbC-StarTrack strategy is compared, the cells labeled in the adult OB following in utero electroporation (IUE) are similar to those drawn by Cajal (Figure 1A–F). Mitral cells (Figure 1(Ae),C) are the first cell type to be born in the rodent OB, between E10 and E13, and with a neurogenic peak around E11 [5]. The axonal projections of mitral cells form the lateral olfactory tract (LOT) and they establish direct contacts with the OC [6,7].

Olfactory interneurons (periglomerular and granule cells) are a diverse group of cells located within the glomerular layer (GL) and granular cell layer (GcL; Figure 1D–E). These interneurons arise from progenitors located within the ganglionic eminences that migrate tangentially to their destination in the OB [8,9]. Neural stem cells (NSCs) in the subventricular zone (SVZ) also give rise to olfactory interneurons during postnatal life, and these progenitors are determined between E13 and E15 [9]. The different kinds of interneurons are generated from embryonic to postnatal stages [10–13], and their temporal origin defines the interneuronal diversity [14]. Glial cells are also widespread in the different layers of the OB, those found in each layer arising from different or the same progenitors (Figure 1F). For example, some astrocytes surrounding a single glomerulus have been shown to be clonally-related [15]. In the embryo, glial progenitor cells are located in the most rostral part of the lateral ventricle (LV), which corresponds to the olfactory ventricle (OV) [15–17]. This complex organization and connectivity of the cells that populate the OB are largely determined during embryonic development [5,18,19], albeit with an additional contribution postnatally [20,21].

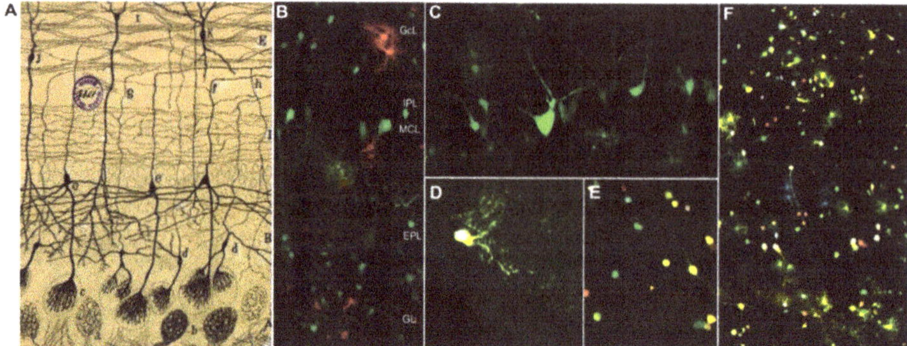

Figure 1. (**A**) Original drawing by Cajal of an olfactory bulb (OB) section from the brain of a perinatal cat [3] showing the glomerular layer (*A*); external plexiform layer (*B*); mitral cell layer (MCL; *C*); internal plexiform layer (*D*); granule cell layer and white matter (*E*); (*a,b*) terminal axonal arborizations of olfactory sensory neurons; (c) dendritic arborizations from tufted (*d*) and mitral cells (e) that form the glomerulus; (*f–h*) axonal projections from tufted and mitral cells); (*I–J*) granule cells; (*K*) short axon cells of the granule cell layer (Cajal Legacy, Instituto Cajal-CSIC, Madrid, Spain). (**B–F**) Adult OB neural cells labeled after in utero electroporation (IUE) of UbC-StarTrack constructs into the E12 mouse embryo lateral ventricle (LV). (**B**) Coronal section of the mouse OB in which UbC-StarTrack labelling shows the different cells that compose the layers described by Cajal. (**C**) Detail of the MCL, with projection neurons and glial cells labeled with UbC-EGFP-StarTrack. Detail of labeled periglomerular (**D**) and granule cells (**E**). (**F**) UbC-StarTrack labeled glia widely spread across the different OB layers. GcL, granular cell layer; IPL, internal plexiform layer; MCL, mitral cell layer; EPL, external plexiform layer; GL, glomerular layer.

To date, different approaches have been used to assess the diversity of OB progenitor pools during development, including the use of fluorescent and lipophilic tracers, viral vectors, immunostaining, and the generation of specific mouse lines. Nevertheless, the heterogeneity of progenitor cells has yet to be fully defined, and more recent single-cell transcriptomic analyses have shed new light on the diversity and potential of progenitor cells [22,23]. Moreover, single-cell lineage tracing revealed the fate potential and lineage progression of some progenitors [24–26].

Here, in order to decipher the heterogeneity of progenitor cells, using UbC-StarTrack lineage tracing approaches under the specific regulation of different promoters, we targeted specific progenitors by IUE to analyze the fate potential of NSCs in the adult brain. The determination of specific cell types in the OB can be influenced by either the molecular profile by their progenitors, the age of the embryo, and/or the location of the labeled progenitors. The data we obtained here confirm that some degree of diversity is present in the pool of OB progenitor cells, highlighting the need of performing further single-cell analyses to define the progenitor cell identities required to generate the complex OB cytoarchitecture. We demonstrate that the origin, fate, and targeting of progenitors must be taken into consideration when studying OB heterogeneity.

2. Materials and Methods

2.1. Mouse Line

C57BL/6 mice were housed at the animal facility of the Cajal Institute. All procedures were carried out in accordance with the guidelines of the European Union on the use and welfare of experimental animals (2010/63/EU) and those of the Spanish Ministry of Agriculture (RD 1201/2005 and L 32/2007). All the experiments were approved by the CSIC Bioethical Committee (PROEX 223/16). The day of visualization of the vaginal plug was considered as embryonic day (E0) and the day of birth as postnatal day (P0). In addition, mice were considered adults from P30 onwards. In all the experiments, a minimum of $n = 3$ animals was considered.

2.2. Vectors

StarTrack constructs were designed as described previously [16,17], and different combinations of StarTrack constructs were used separately to target the different profiles of the progenitor cells. The hyperactive transposase of the PiggyBac system (CMV-hyPBase) was used to generate different vectors in which the expression of the transposases was driven by promoters for NG2, GFAP, and GSX2. The cloning of the different hyPBase constructs was performed by Canvax Biotech, and the source of the promoters is indicated in Table 1. All plasmids were sequenced (Sigma–Aldrich; Saint Louis, MO, USA) to confirm successful cloning. This strategy allowed specific progenitors with active gene expression of these promoters at the time of electroporation to be labeled in order to track their full progeny. Plasmid mixtures contained the twelve UbC-StarTrack floxed constructs, a transposase of the PiggyBac system under the control of the selected specific promoter (either CMV, NG2, GFAP or Gsx2), and the CAG-CreERT2 vector to remove the episomal copies of constructs [17].

Table 1. List of the different plasmids used in the StarTrack approach

Vectors	Promoter	Source	Abbreviation
PiggyBac plasmid	Ubiquitin C	Prof. Bradley	UbC-StarTrack
PiggyBac Transposase	CMV	Prof. Bradley	CMV-hyPBase
	NG2	Kirchoff	NG2-hyPBase
	GFAP	Dr Lundberg	GFAP-hyPBase
	Gsx2	Dr K. Campbell	Gsx2-hyPBase
Cre-recombinase	CAG	Dr C. Cepko	Cre-ERT2

2.3. In Utero Electroporation (IUE) and Tamoxifen Administration

In utero electroporation was performed as described previously [17,27]. Briefly, the selected plasmid mixture was injected into the LV of E12 embryos with a micropipette and using an ultrasound device (VeVo-770; VisualSonics, Toronto, Canada) and then co-electroporated (three animals per experimental group). The embryos were returned to the dam's abdominal cavity, which were then monitored for three days. After birth, all the pups were injected with tamoxifen (Tx, 20 mg/ml dissolved in pre-warmed corn oil: Sigma–Aldrich) to eliminate episomal copies of the plasmids and to achieve heritable and stable labelling of the cell progeny [17]. A single dose of Tx (5 mg/40 gr body weight) was administered intraperitoneally (i.p.) to the litter at P5. The mice were analyzed from P30 onwards.

2.4. Tissue Processing

All mice were analyzed at adult stages, anesthetizing them with an i.p. injection of pentobarbital (Dolethal 40–50 mg/Kg; Vetoquinol, Alcobendas, Madrid) and then perfused with 4% paraformaldehyde (PFA) in 0.1M phosphate buffer (PB). Subsequently, the brain of mice was removed and postfixed for two hours in fresh 4% PFA and then in PB. Coronal vibratome sections of the brains (50 µm) were obtained and mounted onto a glass slide with Mowiol for storage at 4 °C.

2.5. Imaging

The sections were examined under an epifluorescence microscope (Eclipse E600; Nikon Instruments, Melville, NY, USA), equipped with GFP (FF01-473/10), mCherry (FF01-590/20) and Cy5 (FF01-540/15) filters. Images were then acquired on a TCS-SP5 confocal microscope (Leica Microsystems, Wetzlar, Germany) using a 20x objective (Leica), with the wavelength conformation as described previously (Table 2) [17,21]. Confocal laser lines were maximal around 40% in all samples. Maximum projection images were analyzed using LASAF Leica and Fiji software ImageJ (https://imagej.net/Fiji/Downloads). All stitching and contrast adjustments were performed with LasX software (LasX Industries; St Paul, MN, USA) and Photoshop CS5 software (Adobe Inc.; San Jose, CA, USA).

Table 2. Excitation and emission wavelengths for each fluorescent protein reporter

Wavelength (nm)	YFP	mKO	mCerulean	mCherry	mTSapphire	EGFP
Excitation	514		458	561	405	488
Emission	520–535	560–580	468–480	601–620	520–535	498–514

YFP: Yellow fluorescent protein; mKO: Monomeric kusabira orange; EGFP: Enhanced green fluorescent protein.

2.6. Data Analysis

For each experiment, the number of labeled OB cells per section was quantified along the rostrocaudal axis within the OB (32–40 sections per animal). Cells were counted using the manual cell counter plug-in of ImageJ software and the percentage of those cells, located in specific areas, was calculated. The study of 26,685 labeled interneurons in the OB was considered in this approach. For statistics, GraphPad Prism 6.0 (GraphPad, San Diego, CA, USA) was used, and the statistical significance between two groups was assessed with two-tailed unpaired Student's *t*-tests. For multiple comparison study, one-way analysis of variance (ANOVA) was used. A confidence interval of 95% ($p < 0.05$) was used to determine statistically significant values. Critical values of * $p < 0.05$, ** $p < 0.01$, and *** $p < 0.001$ were adopted to determine statistical differences. Graphs were obtained using GraphPad Prism and CorelDRAW Graphic Suite 2018 (Corel Corporation, Ottawa, Canada).

3. Results

3.1. The Fate of OB Cells After Targeting Cell Progenitors at Distinct Ventricular Sites

Using UbC-StarTrack plasmids (Figure 2A), we performed different IUEs at E12 that targeted different ventricular areas (dorsal, ventral, medial) and the most rostral portion of LV, the OV (Figure 2B). Animals were injected with Tx at P5 to remove the episomal copies of the constructs and analyzed at adult stages (from P30 onwards). Rostral IUE, restricted to the rostral OV, labeled glial cells, mitral cells, and some interneurons in the OB (Figure 2C). Interestingly, these glial cells were radially disposed in the different layers of the OB close to the electroporation area. Mitral cells in the mitral cell layer (MCL) were identified through their morphology and the presence of reelin (data not shown). These results indicated that glial and mitral cells originated from progenitor cells located in the most rostral part of the LV at E12. By contrast, when the dorsal, medial, and ventral walls of the LV were targeted, the labeled cells in the OB were periglomerular and granular interneurons, not glial or mitral cells (Figure 2D–I).

After targeting E12 progenitor cells within the dorsal LV, different neural cells were labeled in the adults, spread throughout the corpus callosum and cortex (Figure 2D), although only interneurons were labeled in the OB (Figure 2E). Likewise, ventral electroporation at E12 labeled neurons in the striatum, piriform cortex, and corpus callosum and interneurons in the dorsal cortex (Figure 2F) and the OB (Figure 2G). Dorsal and ventral electroporation mostly labeled interneurons in the GcL, with a few periglomerular cells also labeled. By contrast, medial electroporation labeled cells in the septal area of the telencephalon (Figure 2H), although most cells were located in the GL of the OB (Figure 2I). Finally, IUE of the third ventricle did not label glia or neurons in the OB (data not shown).

In summary, after targeting different ventricular areas at E12, the adult labeled cell-progeny displayed different morphologies at different locations in both OB and forebrain. Thus, the origin of the progenitor cells in specific areas determines their cell fate in the adult telencephalon.

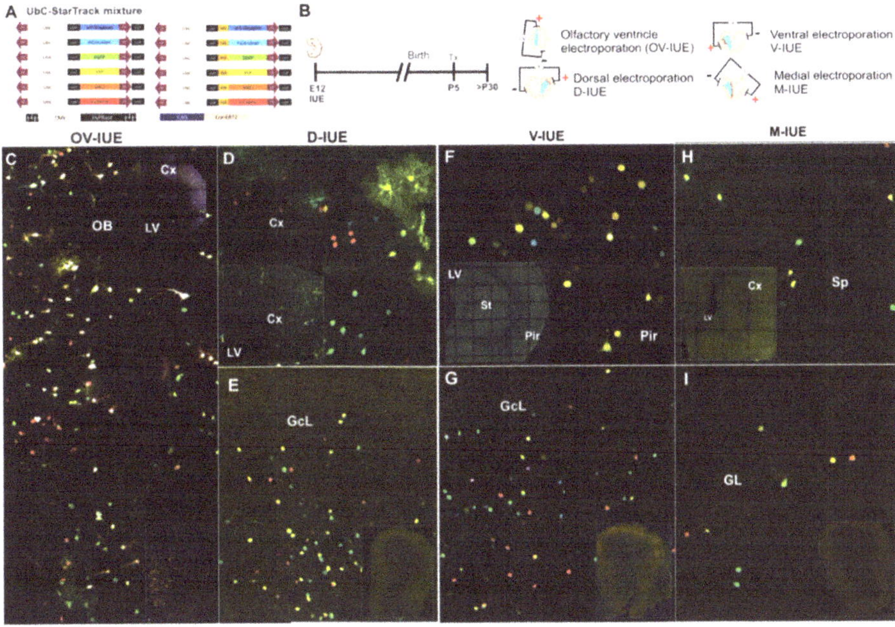

Figure 2. (**A**) Diagram of the UbC-StarTrack vectors, 12 different plasmids encoding six different fluorescent proteins at two different locations, cytoplasmic and nuclear according to the H2B sequence. All vectors were driven by the Ubiquitin C promoter. (**B**) Summary of the IUE procedure, where E12 embryos were injected with UbC-StarTrack mixture and electroporated. After birth Tamoxifen (Tx) was injected at around P5, and the adult tissue was analyzed (>P30). Four different orientations of the electrodes were used for electroporation: olfactory ventricle (OV-IUE), dorsal (D-IUE), ventral (V-IUE), and medial (M-IUE). The red line illustrates the electroporation area. UbC-StarTrack OV-IUE labeled both neurons and glia in the olfactory bulb (OB, **C**). Targeted cells in each lateral ventricular (LV) zone gave rise to different labeled neural cells in the dorsal cortex (**D**), piriform cortex (**F**), and septum (**H**). By contrast, dorsal-, ventral-, and medial- IUE did not produce any labeled glia in the OB; only interneurons were targeted (**E**,**G**,**H**). Dorsal and ventral-IUE targeted progenitors that gave rise to labeled cells in the GcL and eventually, the GL. However, M-IUE produced more labeled cells in the GL. The white squares represent the electroporation area in the telencephalon and OB (**C**–**I**). IUE, in utero electroporation; OB, olfactory bulb; LV, lateral ventricle; Cx, cerebral cortex; Pir, piriform cortex; St, striatum; Sp, septum.

3.2. The Fate of Olfactory Bulb Cells After Targeting Specific Progenitors with StarTrack

We analyzed the fate of progenitor cells using a novel UbC-StarTrack strategy based on the combination of UbC-StarTrack plasmids with different PiggyBac transposases driven by specific promoters. This strategy drives the integration of the plasmids exclusively into the progenitors that express the specific promoters chosen at the time of electroporation (Figure 3A). As such, we specifically targeted NSCs using the CMV, NG2, Gsh-2, and GFAP promoters. First, the UbC-StarTrack and CMV-transposase (CMV-hyPBase: Figure 3B) incorporated copies of the plasmids ubiquitously, labelling all the progenitor cells and their progeny. Subsequently, the PiggyBac transposase encoding the NG2 promoter (NG2-hyPBase: Figure 3C) was used to target only those progenitor cells with an active NG2 promoter, integrating copies of the plasmids and labeling their progeny. In another approach, the PiggyBac transposase was driven by the subpallial promoter Gsh-2 promoter (Gsh2-hyPBase: Figure 3D) to only label the progenitors located in the ganglionic eminences at early developmental stages and consequently, their adult cell progeny. Finally, the GFAP promoter was incorporated into a transposase (GFAP-hyPBase) and co-electroporated with UbC-StarTrack to label GFAP-progenitor cells (Figure 3E). All these IUEs were directed at the dorso-lateral ventricle walls, except for the Gsh2-hyPBase, which was ventrally orientated. As a result of these manipulations, all the labeled cells in the OB corresponded to interneurons situated in the GL and GcL, with no glial cells or projection neurons. This comparative analysis of the different StarTrack vectors involved 12 animals ($n = 3$ for each transposase driven by a different promoter) and the study of 26,685 labeled interneurons in the OB, of which 12,236 were generated by progenitors expressing NG2; 8308 were from CMV progenitors; 5035 were from progenitors electroporated with the Gsh-2 transposase; and only 1,106 cells were from progenitor cells expressing GFAP. However, no significant differences were evident for each construct in terms of the average of labeled interneurons in the OB (Figure 3F).

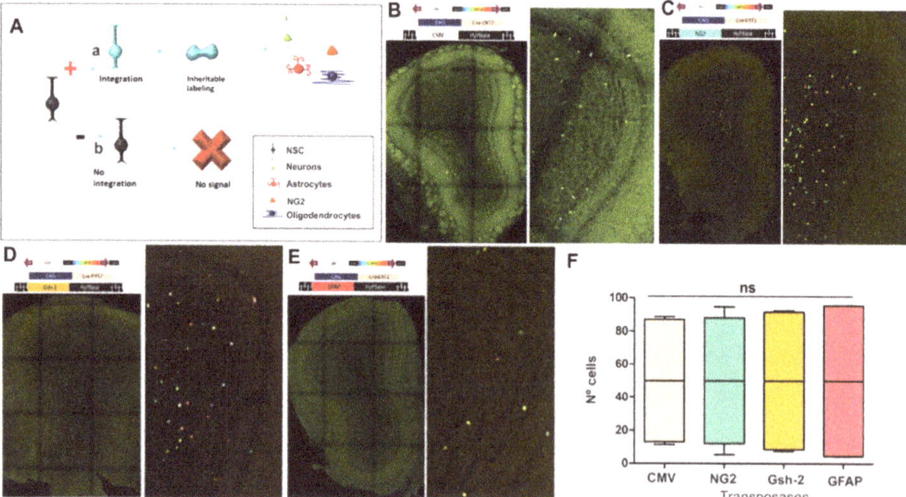

Figure 3. Diagram of the UbC-StarTrack strategy based on transposase promoter expression (**A**). The concept that is focused in the transposase only integrates copies of the UbC-StarTrack vectors into progenitor cells with the corresponding promoter active, labeling all their progeny (**a**). Progenitor cells with the inactive promoter do not integrate copies into the NSCs (**b**). For these experiments, the CMV, NG2, Gsh-2, and GFAP promoters were chosen to target specific NSCs. The first strategy with CMV-hyPBase labeled OB interneurons in the different layers (**B**). The NG2 progeny labeled cells in the GcL and GL (**C**), resembling the Gsh-2 progeny (**D**). GFAP progenitors gave rise to granular cells and periglomerular cells (**E**). GFAP progenitors produced fewer labeled cells in the OB than the other vectors (**F**). All data were normalized; the box plot represents the percentage of labeled cells after targeting each set of progenitors with a specific transposase (whiskers represent 5th/95th percentile, horizontal line displays the median of the data; $n = 3$ for each transposase). Data showed no statistically significant difference between groups (ns). CMV-progenitors are shown in soft pink; NG2-progenitors in blue; Gsh-2 in yellow; GFAP-progenitors in red.

Therefore, these results indicate that the pool of progenitor cells committed to give rise to OB interneurons was quite heterogeneous. Accordingly, NG2 and CMV progenitors at E12 produced a larger proportion of adult OB cells compared to those produced from progenitors expressing GFAP.

3.3. Diversity of Olfactory Bulb Interneurons in Relation to Progenitor Cell Identity

Considering the molecular profile of specific NCSs, we studied the differences between the interneurons generated by the different pools of progenitor cells. The UbC-StarTrack plasmids and the CreERT2 vector were injected along with one of the specific transposases (CMV, NG2, GFAP or Gsh-2) (Figure 4A–E). The distribution of the labeled cells in the adult OB was analyzed and correlated with their progenitor cell profile. All the labeled cells were interneurons, periglomerular, and granular cells, even though some immature cells were found close to the subependymal zone (data not shown). Of the cells labeled by the transposase driven by the CMV promoter, 14% were located in the GL, while 86% were located in the GcL (Figure 4B). When NG2 and Gsh2 drove transposase expression, a similar proportion of cells was found in the GcL (88% NG2, 89% Gsh-2) and GL (12% NG2, 11% Gsh-2: Figure 4C,D). However, after targeting the GFAP progenitors, the cell-derived progeny was preferentially sited within the GcL (93%) rather than in the GL (7%: Figure 4E). Besides, these GFAP-progenitors are committed preferably to external areas of GL compared with those that express other promoters. In summary, progenitor cells were committed to preferentially generate granule cells more than periglomerular cells (Figure 4F). Otherwise, there were no significant differences between the distinct types of progenitor cells committed to generate periglomerular and granular cells (Figure 4G).

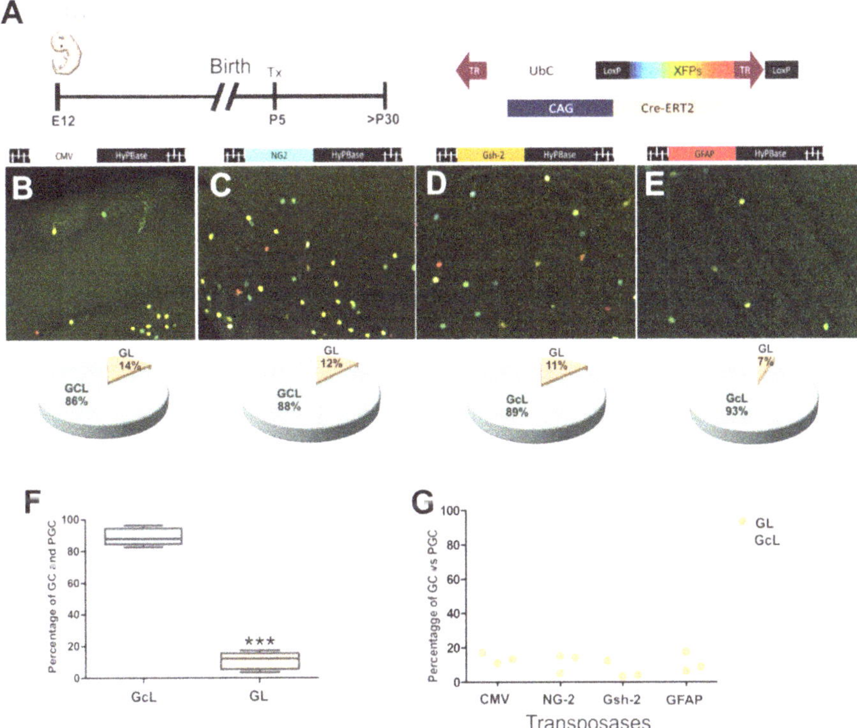

Figure 4. IUE at E12 with the UbC-StarTrack constructs (**A**) and CAG-Cre-recombinase, along with the transposase (**B–E**). All animals were injected with Tx at P5 to remove the episomal copies of the

UbC-constructs, and the brain was analysed from P30 onwards. Of the cells produced by CMV-progenitors, 86% were in the GcL and 14% in the GL (**B**). The NG2-progenitors produced 88% GcL cells and 12% GL cells (**C**), similar to the Gsh-2 progenitors (**D**), while GFAP progenitors gave rise to only 7% of GL cells (**E**). Nevertheless, more labeled cells were located in the GcL (green) than in the GL (soft pink). The box plot represents the percentage of labeled cells after targeting each set of progenitors with a specific transposase, and the line displays the mean of the data (box and whisker $5^{th}/95^{th}$ percentile plot). A confidence interval of 95% ($p < 0.05$) was used to determine statistically significant values (***$p < 0.001$). (**F**). Pallial and subpallial electroporations into the LV produced more cells in the GcL than GL (three animals were analyzed per experiment: CMV; NG2; Gsh-2; GFAP ($n = 12$). Data are shown as average data points (**G**).

Accordingly, these results suggest that E12 progenitors in dorsal and ventral LV produce more granule cells than periglomerular cells, and this cell fate is independent of the molecular profile of NSC.

In brief, these results summarize the importance of the genes targeted, the location, and the identity of progenitor cells when studying the heterogeneity of specific populations, in this case, adult OB cells (Figure 5). The data obtained open the window for further transcriptomic and clonal studies of these populations in order to define the heterogeneous lineages present in the adult brain.

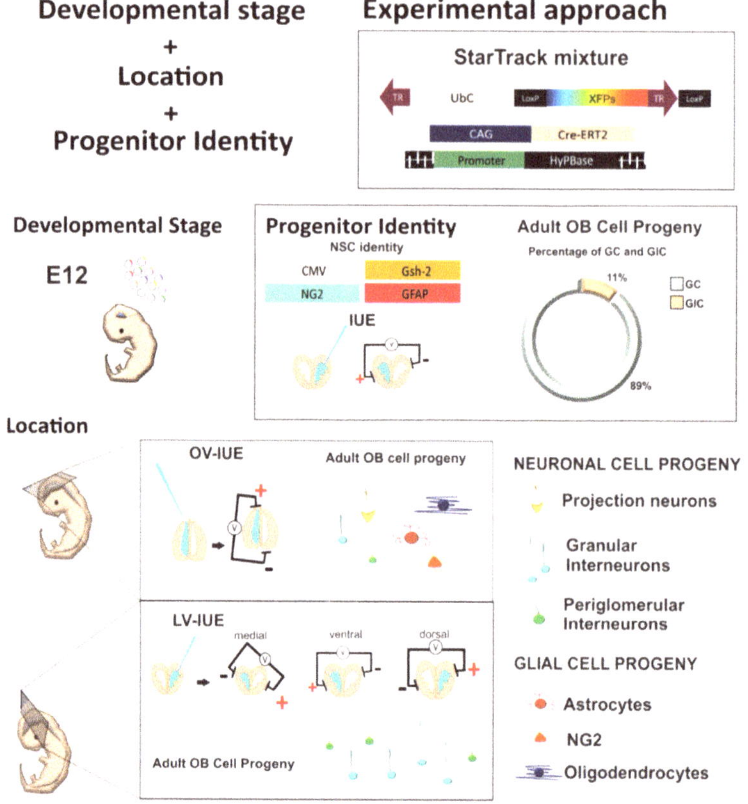

Figure 5. Summary of the importance of age, location, and cell identity to reveal the heterogeneity of NSCs after targeting with the UbC-StarTrack mixture. E12 progenitor cells lining the lateral ventricles can give rise to different neural cell types in the OB depending on their location in the neurogenic/gliogenic niches. The identity of the progenitor is crucial to define the potential fate of the progenitor cells.

4. Discussion

In this study, a novel StarTrack approach was adopted to address the ontogeny of different cell types in the adult rodent OB, taking into account the identity of their progenitor cells and their location in the LV. We focused on the genetic profile of progenitor cells in order to define the heterogeneity of the NSCs that give rise to neural cells in the adult OB and the origin of these cells. Specific progenitor cells were targeted at E12 to track their adult cell progeny in the OB. StarTrack is a powerful tool to examine the fate and the clonal relationship between cells derived from specific progenitors in vivo, under both physiological and pathological conditions [15–17,21,25,27–33]. This tool also allows progenitor cells to be tracked in vivo, avoiding genetic manipulation of the animals or viral injections. In our particular case, NSCs in the LV were targeted by IUE, avoiding targeting progenitor cells at other sites in which these promoters may be active, such as pericytes in the case of the NG2 promoter [34–36]. NG2-hyPBase-driven integration of the StarTrack mix via IUE targets those neural progenitors in the LV with an active NG2-promoter, thereby limiting the developmental spatio-temporal parameters of the study (reviewed by Shimogory and Ogawa [37]).

The heterogeneity of NSC pools is an issue that has yet to be resolved [20,38]. The past two decades have witnessed an accumulation of abundant evidence regarding such heterogeneity, not only in the OB, but also in the dorsal cortex [39], hippocampus [40], and cerebellum [30]. Moreover, recent data show the bipotent capacity of postnatal NSCs to generate OB interneurons and glia in the cortex and striatum [21]. NSCs generating OB cells follow highly patterned and complex behavior during embryogenesis [41,42], and at post-natal stages [13,41]. Single cell analysis provides new insight into the development of the forebrain and the changes it undergoes in the adult. Clonal and transcriptomic analyses make it possible to explore the huge heterogeneity of neural cells [24,32,43,44]. Indeed, the heterogeneity of cortical progenitor cells was recently reported, showing the cell lineage of cortical pyramidal cells restricted to either the deep or superficial layers [39], suggesting that the heterogeneity of neocortical progenitor cells is greater than that previously thought [45–47]. Currently, the extent of NSC heterogeneity still remains to be defined as such, and further studies and improved methods will be required to fully specify the cell diversity in the telencephalon.

Thus, it is relevant to target certain sites in order to study specific cell progeny. In particular, the origin of glial cells is not as well understood as that of neurons, and it has been suggested that glial cells travel from the LV to OB via the RMS [48]. The data presented here show that glial and mitral cells progenitors are located in the OV, as previously reported [49]. Nevertheless, it is important to consider those cells in the adult OB that were not targeted by our strategy due to the location of their progenitors, as the ensheathing cells migrating from the olfactory placode to the olfactory nerve layer of the OB [50]. While our data showed the importance of embryonic progenitor location more than the cell identity at E12, neonatal progenitor microdomains may also exist in which specific OB cells originate [13,41].

In summary, the results presented here open a window to explore the genomic profile of neural progenitor cells in more detail. Moreover, they highlight the importance of further studies at the single-cell level to define the heterogeneity of NSC populations and their progeny in the OB. Such studies will help us to better understand this complex brain structure.

Author Contributions: Conceptualization, L.L.-M.; methodology, R.S.-G., M.F.-O., A.C.O.-S.; formal analysis, R.S.-G.; investigation, R.S.-G.; writing—original draft preparation, L.L.-M., R.S.-G.; writing—review and editing, L.L.-M., R.S.-G., M.F.-O., A.C.O.-S; project administration, L.L.-M.; funding acquisition, L.L.-M. All authors have read and agreed to the published version of the manuscript.

Funding: This research was funded by the Spanish Ministry of Economy and Competitivity (MINECO), grant number BFU2016-75207-R.

Acknowledgments: We are very grateful to both the Imaging and Microscopy Facility at the Instituto Cajal and the Microscopy and Image Analysis Services at the National Hospital for Paraplegics, Toledo.

Conflicts of Interest: The authors declare no conflict of interest. The funders had no role in the design of the study; in the collection, analyses, or interpretation of data; in the writing of the manuscript, or in the decision to publish the results.

References

1. Ramón y Cajal, S. Origen y terminación de las fibras nerviosas olfatorias. *Gaceta Sanitaria de Barcelona* **1890**, *3*, 1–20. Available online: https://digital.csic.es/bitstream/10261/158702/1/Cap%C3%ADtulo43-T2-2%C2%AA%20parte.pdf (accessed on 25 December 2019).
2. Blanes, T. Sobre algunos puntos dudosos de la estructura del bulbo olfatorio. *Rev. Trimest. Microgr.* **1898**, *3*, 99–127.
3. Ramón y Cajal, S. Estudios sobre la corteza cerebral humana IV. Estructura de la corteza cerebral olfativa del hombre y mamíferos. *Trab. Lab. Investig. Biol.* **1901**, *1*, 1–140.
4. Figueres-Oñate, M.; Gutiérrez, Y.; López-Mascaraque, L. Unraveling Cajal's view of the olfactory system. *Front. Neuroanat.* **2014**, *8*, 55. [CrossRef]
5. Blanchart, A.; de Carlos, J.A.; López-Mascaraque, L. Time frame of mitral cell development. *J. Comp. Neurol.* **2006**, *543*, 529–554. [CrossRef]
6. Allison, A.C. The secondary olfactory areas in the human brain. *J. Anat.* **1954**, *88*, 481–488.
7. Price, J.L. An autoradiographic study of complementary laminar patterns of termination of afferent fibers to the olfactory cortex. *J. Comp. Neurol.* **1973**, *150*, 87–108. [CrossRef]
8. Kriegstein, A.; Alvarez-Buylla, A. The glial nature of embryonic and adult neural stem cells. *Annu. Rev. Neurosci.* **2009**, *32*, 149–184. [CrossRef]
9. Fuentealba, L.C.; Rompani, S.B.; Parraguez, J.I.; Obernier, K.; Romero, R.; Cepko, C.L.; Alvarez-Buylla, A. Embryonic origin of postnatal neural stem cells. *Cell* **2015**, *161*, 1644–1655. [CrossRef]
10. Ventura, R.E.; Goldman, J.E. Dorsal radial glia generates olfactory bulb interneurons in the postnatal murine brain. *J. Neurosci.* **2007**, *27*, 4297–4302. [CrossRef]
11. Merkle, F.T.; Mirzadeh, Z.; Alvarez-Buylla, A. Mosaic organization of neural stem cells in the adult brain. *Science* **2007**, *317*, 381–384. [CrossRef]
12. Batista-Brito, R.; Close, J.; Machold, R.; Fishell, G. The distinct temporal origins of olfactory bulb interneuron subtypes. *J. Neurosci.* **2008**, *28*, 3966–3975. [CrossRef]
13. Fiorelli, R.; Azim, K.; Fischer, B.; Raineteau, O. Adding a spatial dimension to postnatal ventricular-subventricular zone neurogenesis. *Development* **2015**, *142*, 2109–2120. [CrossRef]
14. De Marchis, S.; Bovetti, S.; Carletti, B.; Hsieh, Y.C.; Garzotto, D.; Peretto, P.; Fasolo, A.; Puche, A.C.; Rossi, F. Generation of distinct types of periglomerular olfactory bulb interneurons during development and in adult mice: Implication for intrinsic properties of the subventricular zone progenitor population. *J. Neurosci.* **2007**, *27*, 657–664. [CrossRef]
15. García-Marqués, J.; López-Mascaraque, L. Clonal mapping of astrocytes in the olfactory bulb and rostral migratory stream. *Cereb. Cortex* **2017**, *27*, 2195–2209. [CrossRef]
16. García-Marqués, J.; López-Mascaraque, L. Clonal identity determines astrocyte cortical heterogeneity. *Cereb. Cortex* **2013**, *23*, 1463–1472. [CrossRef]
17. Figueres-Oñate, M.; García-Marqués, J.; López-Mascaraque, L. UbC-StarTrack, a clonal method to target the entire progeny of individual progenitors. *Sci. Rep.* **2016**, *6*, 33896. [CrossRef]
18. Jiménez, D.; García, C.; de Castro, F.; Chédotal, A.; Sotelo, C.; De Carlos, J.A.; Valverde, F.; López-Mascaraque, L. Evidence for intrinsic development of olfactory structures in *Pax-6* mutant mice. *J. Comp. Neurol.* **2000**, *428*, 511–526. [CrossRef]
19. López-Mascaraque, L.; de Castro, F. The olfactory bulb as an independent developmental domain. *Cell Death Differ.* **2002**, *9*, 1279–1286. [CrossRef]
20. Obernier, K.; Alvarez-Buylla, A. Neural stem cells: Origin heterogeneity and regulation in the adult mammalian brain. *Development* **2019**, *146*, dev156059. [CrossRef]
21. Figueres-Oñate, M.; Sánchez-Villalón, M.; Sánchez-González, R.; López-Mascaraque, L. Lineage tracing and clonal cell dynamics of postnatal progenitor cells *in vivo*. *Stem Cell Rep.* **2019**, *13*, 700–712. [CrossRef]

22. Codega, P.; Silva-Vargas, V.; Paul, A.; Maldonado-Soto, A.R.; DeLeo, A.M.; Pastrana, E.; Doetsch, F. Prospective identification and purification of quiescent adult neural stem cells from their in vivo niche. *Neuron* **2014**, *82*, 545–559. [CrossRef] [PubMed]
23. Tiwari, N.; Pataskar, A.; Péron, S.; Thakurela, S.; Sahu, S.K.; Figueres-Oñate, M.; Berninger, B. Stage-specific transcription factors drive Astrogliogenesis by remodeling gene regulatory landscapes. *Cell Stem Cell* **2018**, *23*, 557–571.e8. [CrossRef] [PubMed]
24. Calzolari, F.; Michel, J.; Baumgart, E.V.; Theis, F.; Gotz, M.; Ninkovic, J. Fast clonal expansion and limited neural stem cell self-renewal in the adult subependymal zone. *Nat. Neurosci.* **2015**, *18*, 490–492. [CrossRef] [PubMed]
25. Bribián, A.; Figueres-Oñate, M.; Martín-López, E.; López-Mascaraque, L. Decoding astrocyte heterogeneity: New tools for clonal analysis. *Neuroscience* **2016**, *323*, 10–19. [CrossRef] [PubMed]
26. Ma, J.; Shen, Z.; Yu, Y.C.; Shi, S.H. Neural lineage tracing in the mammalian brain. *Curr. Opin. Neurobiol.* **2018**, *50*, 7–16. [CrossRef] [PubMed]
27. Figueres-Oñate, M.; García-Marqués, J.; Pedraza, M.; De Carlos, J.A.; López-Mascaraque, L. Spatiotemporal analyses of neural lineages after embryonic and postnatal progenitor targeting combining different reporters. *Front. Neurosci.* **2015**, *9*, 87. [CrossRef] [PubMed]
28. Martín-López, E.; García-Marques, J.; Núñez-Llaves, R.; López- Mascaraque, L. Clonal astrocytic response to cortical injury. *PLoS ONE* **2013**, *8*, e74039. [CrossRef] [PubMed]
29. Parmigiani, E.; Leto, K.; Rolando, C.; Figueres-Oñate, M.; López-Mascaraque, L.; Buffo, A.; Rossi, F. Heterogeneity and bipotency of astroglial-like cerebellar progenitors along the interneuron and glial lineages. *J. Neurosci.* **2015**, *35*, 7388–7402. [CrossRef]
30. Cerrato, V.; Parmigiani, E.; Betizeau, M.; Aprato, J.; Nanavaty, I.; Berchialla, P.; Luzzati, F.; de'Sperati, C.; López-Mascaraque, L.; Buffo, A. Multiple origins and modularity in the spatiotemporal emergence of cerebellar astrocyte heterogeneity. *PLoS Biol.* **2018**, *16*, e2005513. [CrossRef]
31. Bribián, A.; Pérez-Cerda, F.; Matute, C.; López-Mascaraque, L. Clonal glial response in a multiple sclerosis mouse model. *Front. Cell. Neurosci.* **2018**, *23*, 375. [CrossRef]
32. Redmond, S.A.; Figueres-Oñate, M.; Obernier, K.; Nascimento, M.A.; Parraguez, J.I.; López-Mascaraque, L.; Fuentealba, L.C.; Alvarez-Buylla, A. Development of ependymal and postnatal neural stem cells and their origin from a common embryonic progenitor. *Cell. Rep.* **2019**, *27*, 429–441. [CrossRef] [PubMed]
33. Gutiérrez, Y.; García-Marqués, J.; Liu, X.; Fortes-Marco, L.; Sánchez-González, R.; Giaume, C.; López-Mascaraque, L. Sibling astrocytes share preferential coupling via gap junctions. *Glia* **2019**, *67*, 1852–1858. [CrossRef] [PubMed]
34. Nishiyama, A.; Komitova, M.; Suzuki, R.; Zhu, X. Polydendrocytes (NG2 cells): Multifunctional cells with lineage plasticity. *Nat. Rev. Neurosci.* **2009**, *10*, 9–22. [CrossRef] [PubMed]
35. Huang, W.; Zhao, N.; Bai, X.; Karram, K.; Trotter, J.; Goebbels, S.; Scheller, A.; Kirchhoff, F. Novel NG2-CreERT2 knock-in mice demonstrate heterogeneous differentiation potential of NG2 glia during development. *Glia* **2014**, *62*, 896–913. [CrossRef] [PubMed]
36. Huang, W.; Guo, Q.; Bai, X.; Scheller, A.; Kirchhoff, F. Early embryonic NG2 glia are exclusively gliogenic and do not generate neurons in the brain. *Glia* **2019**, *67*, 1094–1103. [CrossRef]
37. Shimogori, T.; Ogawa, M. Gene application with *in utero* electroporation in mouse embryonic brain. *Develop. Growth Differ.* **2008**, *50*, 499–506. [CrossRef]
38. Dimou, L.; Götz, M. Glial Cells as Progenitors and stem cells: New roles in the healthy and diseased brain. *Physiol. Rev.* **2014**, *94*, 709–737. [CrossRef]
39. Llorca, A.; Ciceri, G.; Beattie, R.; Wong, F.K.; Diana, G.; Serafeimidou-Pouliou, E.; Fernández-Otero, M.; Streicher, C.; Arnold, S.J.; Meyer, M.; et al. A stochastic framework of neurogenesis underlies the assembly of neocortical cytoarchitecture. *eLife* **2019**, *8*, e51381. [CrossRef]
40. Gebara, E.; Bonaguidi, M.A.; Beckervordersandforth, R.; Sultan, S.; Udry, F.; Gijs, P.J.; Lie, D.C.; Ming, G.-L.; Song, H.; Toni, N. Heterogeneity of radial glia-like cells in the adult hippocampus. *Stem Cells* **2016**, *34*, 997–1010. [CrossRef]
41. Merkle, F.T.; Fuentealba, L.C.; Sanders, T.A.; Magno, L.; Kessaris, N.; Alvarez-Buylla, A. Adult neural stem cells in distinct microdomains generate previously unknown interneuron types. *Nat. Neurosci.* **2014**, *17*, 207–214. [CrossRef]

42. Sánchez-Guardado, L.; Lois, C. Lineage does not regulate the sensory synaptic input of projection neurons in the mouse olfactory bulb. *eLife* **2019**, *27*, 8. [CrossRef]
43. Rushing, G.V.; Bollig, M.K.; Ihrie, R.A. Heterogeneity of neural stem cells in the ventricular–subventricular zone. *Adv. Exp. Med. Biol.* **2019**, *1169*, 1–30. [CrossRef] [PubMed]
44. Marcy, G.; Raineteau, O. Contributions of single-cell approaches for probing heterogeneity and dynamics of neural progenitors throughout life. *Stem Cells* **2019**, *37*, 1381–1388. [CrossRef]
45. Gao, P.; Postiglione, M.P.; Krieger, T.G.; Hernandez, L.; Wang, C.; Han, Z.; Streicher, C.; Papusheva, E.; Insolera, R.; Chugh, K.; et al. Deterministic progenitor behavior and unitary production of neurons in the neocortex. *Cell* **2014**, *159*, 775–788. [CrossRef] [PubMed]
46. García-Moreno, F.; Vasistha, N.A.; Begbie, J.; Molnár, Z. CLoNe is a new method to target single progenitors and study their progeny in mouse and chick. *Development* **2014**, *141*, 1589–1598. [CrossRef]
47. Gil-Sanz, C.; Espinosa, A.; Fregoso, S.P.; Bluske, K.K.; Cunningham, C.L.; Martinez-Garay, I.; Zeng, H.; Franco, S.J.; Müller, U. Lineage tracing using Cux2-Cre and Cux2-CreERT2 mice. *Neuron* **2015**, *86*, 1091–1099. [CrossRef]
48. Aguirre, A.A.; Gallo, V. Postnatal neurogenesis and gliogenesis in the olfactory bulb from NG2-expressing progenitors of the subventricular zone. *J. Neurosci.* **2004**, *24*, 10530–10541. [CrossRef]
49. Suzuki, S.O.; Goldman, J.E. Multiple cell populations in the early postnatal subventricular zone take distinct migratory pathways: A dynamic study of glial and neuronal progenitor migration. *J. Neurosci.* **2003**, *23*, 4240–4250. [CrossRef]
50. Blanchart, A.; Martín-López, E.; De Carlos, J.A.; López- Mascaraque, L. Peripheral contributions to olfactory bulb cell populations (migrations towards the olfactory bulb). *Glia* **2011**, *59*, 278–292. [CrossRef]

Review

Sustentacular Cell Enwrapment of Olfactory Receptor Neuronal Dendrites: An Update

Fengyi Liang

Department of Anatomy, Yong Loo Lin School of Medicine, National University of Singapore, 4 Medical Drive, Singapore 117594, Singapore; antlfy@nus.edu.sg; Tel.: +65-6516-1936

Received: 2 March 2020; Accepted: 27 April 2020; Published: 30 April 2020

Abstract: The pseudostratified olfactory epithelium (OE) may histologically appear relatively simple, but the cytological relations among its cell types, especially those between olfactory receptor neurons (ORNs) and olfactory sustentacular cells (OSCs), prove more complex and variable than previously believed. Adding to the complexity is the short lifespan, persistent neurogenesis, and continuous rewiring of the ORNs. Contrary to the common belief that ORN dendrites are mostly positioned between OSCs, recent findings indicate a sustentacular cell enwrapped configuration for a majority of mature ORN dendrites at the superficial layer of the OE. After vertically sprouting out from the borderlines between OSCs, most of the immature ORN dendrites undergo a process of sideways migration and terminal maturation to become completely invaginated into and enwrapped by OSCs. Trailing the course of the dendritic sideways migration is the mesodendrite (mesentery of the enwrapped dendrite) made of closely apposed, cell junction connected plasma membrane layers of neighboring folds of the host sustentacular cell. Only a minority of the mature ORN dendrites at the OE apical surface are found at the borderlines between OSCs (unwrapped). Below I give a brief update on the cytoarchitectonic relations between the ORNs and OSCs of the OE. Emphasis is placed on the enwrapment of ORN dendrites by OSCs, on the sideways migration of immature ORN dendrites after emerging from the OE surface, and on the terminal maturation of the ORNs. Functional implications of ORN dendrite enwrapment and a comparison with myelination or Remak's bundling of axons or axodendrites in the central and peripheral nervous system are also discussed.

Keywords: olfactory receptor neuron (ORN); dendrite; enwrapment; olfactory sustentacular cell (OSC)

---

1. Introduction

The main olfactory epithelium (OE) in mammals consists of a relatively simple pseudostratified epithelial cell layer lining the superior/dorsal part of the nasal cavity. Its basic cytological features have been known since at least the middle of the 19th century [1,2]. Apart from the bipolar olfactory receptor neurons (ORNs), the OE also comprises olfactory sustentacular cells (OSCs), horizontal and globose basal cells, ductal cells of Bowman's glands, and sporadic microvillar cells [3–11].

The ORNs are undoubtedly the OE's parenchymal cells responsible for olfactory reception and transduction. These bipolar neuronal cells are directly exposed at the dendritic end to the nasal mucus and potentially harmful agents or microorganisms in the ambient air of the nasal cavity. At the axonal pole, the ORNs are synaptically connected to the olfactory bulb of the central nervous system (CNS) [6,8]. According to their expression of G-protein-coupled odorant receptors, ORNs in the OE could be differentiated into hundreds of subsets, with each subset usually showing only one phenotype of odorant receptor proteins; axons from the same ORN subset converge to selectively project to the same one or a few glomeruli of the olfactory bulb [12,13]. The OSCs are believed to be partly epithelial and partly glial, functioning as major physical, metabolic, secretory, absorptive, phagocytic, and diverse other supports for the ORNs and the OE overall [3,4,10,11,14–17]. The globose and horizontal basal cells represent essentially the precursor or stem cells of the OE that could give rise to other OE cell types, especially ORNs that have a relatively short lifespan of only a few weeks, and therefore are continuously replaced throughout life of the organism [5,18–21]. The nature and roles of the OE microvillar cells remain unclear [7,8], but recent findings point toward modulatory and maintenance functions for these cells in olfactory reception, ORN apoptosis and regeneration, or in OE aging [22–25].

2. Early Cytological and Cytoarchitectonic Studies of the Olfactory Epithelium

The OE serves both epithelial and special sense (olfactory) functions. As an epithelium, it protects and separates deeper structures from the air of the nasal cavity. As a special sense organ, the OE is the site of olfactory reception and signal transduction, and ORN axonal projection to the olfactory bulb transmits olfaction signals to the CNS. These dual roles of the OE are structurally subserved not only by the OE cell types, but also by the cytoarchitectonic organization and interrelations among the cells, especially those between the ORNs and OSCs. Indeed, the study of OE cytology and cytoarchitecture has a long history in various vertebrate species ranging from fish to man. Earlier literature concerning this topic has been extensively reviewed [2,26]. Before the 1970s, it was generally assumed that mammalian ORN dendrites were located at the borderlines between OSCs [2,26], as typically illustrated in Figure 1.

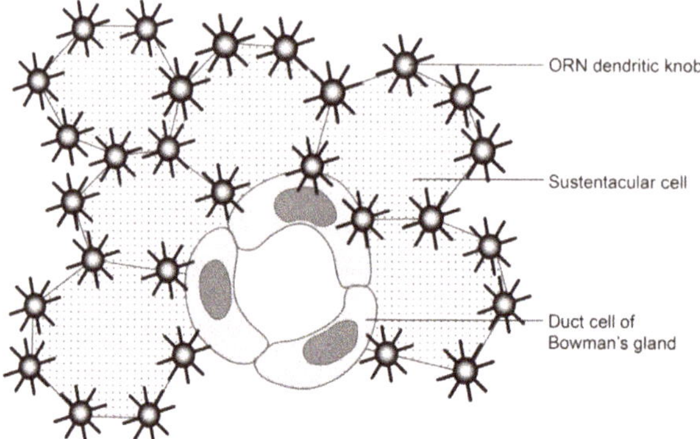

Figure 1. A schematic diagram illustrating major cell types and their relations on the tangential view of the luminal surface of olfactory epithelium of the rabbit. Re-drawn with permits from [2].

Breipohl et al. [3] first reported "a few" sustentacular cell-enclosed ORN dendrites, apart from the "normal" majority of ORN dendrites between OSCs in the OE of the mouse and goldfish. The enclosed ORN dendrites in the goldfish occasionally showed spiral-shaped casing around. In the mouse OE, several ORN dendrites appeared enclosed in one sustentacular cell apical process. The authors further suggested similarities among sustentacular cell enclosure of ORN dendrites, myelination of neuronal axons by Schwann cells or oligodendroglia, and surrounding of optical receptor cell processes by retinal pigment epithelial cells [3].

These findings by Breipohl et al. [3] were subsequently confirmed in the human and rat OE [9,27,28]. By using scanning and transmission electron microscopy, Morrison and Costanzo [27] stated that human ORN cell bodies and dendrites were partially surrounded by OSCs, and ORN axons were surrounded by cellular extensions or sleeve-like processes of OSCs at the basal layer of the OE. Similarly, the scanning electron microscopic study of the rat OE by Nomura and coworkers [28] described groups of ORN cell bodies aligned along vertical columns and roughly incompletely wrapped by flat processes of OSCs. Mature ORN dendritic shafts were often loosely incompletely invested by plicate processes or longitudinal folds of OSCs, and immature ORNs at the basal layer of the OE were also found to be partly enclosed by foot processes of OSCs. Dendrites of immature ORNs, however, were usually independent of sustentacular cell investment. The investment of ORN dendrites by OSCs was suggested to serve a function for micro-electrical isolation, but not so much for guiding initial growth of immature ORN dendrites [28]. More recently, in the study of cell junctions in the OE and olfactory fila, Steinke and colleagues [9] also clearly demonstrated the existence of ORN dendrites embedded into OSCs, in addition to dendrites situated between two or several adjacent supporting cells.

In spite of these and other works, up to the present day it is still generally believed that an overwhelming majority of ORN dendrites in the OE are located between OSCs [29,30] (Figure 2), and the wrapping of ORN dendrites by OSCs, if recognized at all, has been characterized as occasional and partial [31–34].

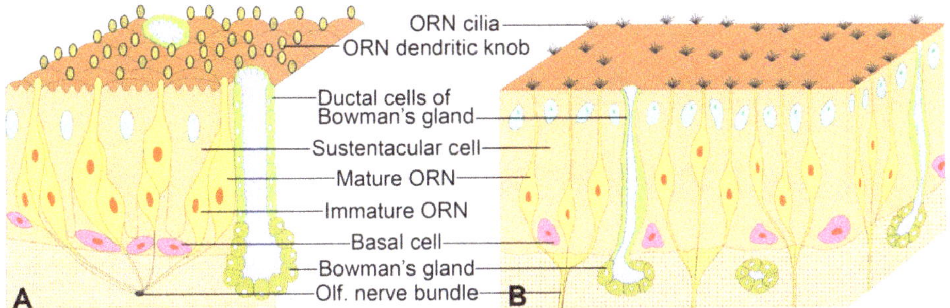

Figure 2. Schematic illustrations of the olfactory epithelium cytology as described in some of the recent publications [29,30]. Both illustrations clearly assumed olfactory receptor neuron (ORN) dendrite positions between olfactory sustentacular cells (OSCs). (**A**) Re-drawn and modified with permits from [29] and (**B**) re-drawn from [30] with permits.

In the following section, I would focus on some of the recent findings on the cytoarchitectonic relations of ORN dendrites and OSCs of adult rat OE. Although the relations in many other vertebrates including man remain largely unclear, we have good reasons to believe that these findings in the rat most likely apply, at least partially, to the OE of many other species, given the aforementioned previous findings, and the fact that the basic cytology and cytoarchitectonic organization of the OE are rather conserved all through the vertebrates [2,26].

3. A Majority of Olfactory Receptor Neuron Dendrites are Enwrapped by Sustentacular Cells

Under a confocal or electron microscope, the cytoarchitectonic relations between cross-sectioned ORN dendrites and OSCs or other cells could be readily visualized on tangential or oblique histological sections of the OE superficial layer stained by ZO1 (zonula occludens-1, a tight junction protein) immunohistochemistry or actin cytoskeleton histochemistry. Apart from unwrapped ORN dendrites (examples being labeled by "u" in Figure 3) located at the borderlines between OSCs or occasionally between an OSC and a microvillar cell, there exist ORN dendrites completely-wrapped (enwrapped) or partially-wrapped by OSCs. A cross-sectioned enwrapped dendrite (examples being labeled by "e" in Figure 3) appears circular, completely invaginated into a vertical passage barrel inside a single OSC, and linked to the borderline between the host and a neighboring OSC by a mesentery of the enwrapped dendrite (mesodendrite) (Figure 3, arrowheads). A partially-wrapped dendrite (one example being labeled by "p" in Figure 3) is largely invaginated into a deep vertical groove on the side of an OSC, but a small part (less than 1/4) of its surface remains directly linked to the inter-sustentacular borderline. Therefore, partially-wrapped dendrites display no observable mesodendrites.

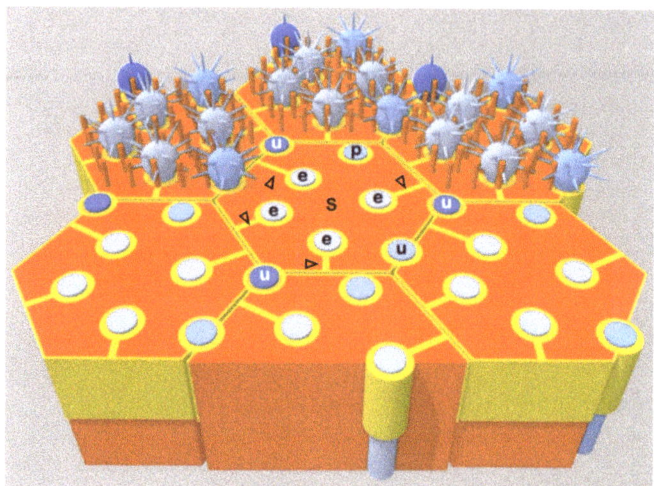

Figure 3. A three-dimensional diagram to show the rat olfactory epithelium (OE) apical layer. Six and a half OSCs (S, orange) and related ORN dendrites are shown, three with intact OSC microvilli, related dendritic knobs, and cilia (top row), and the rest having the OSC microvilli and related dendritic knobs removed to better illustrate relations among the cells. Newly emerged ORN dendrites are immature (highly immunoreactive for class-III β-tubulin) and located at the borderlines between OSC apices (unwrapped, three of them being labeled by "u" in white). Along the course of further maturation (as marked by progressively weakening class-III β-tubulin immunoreactivity), most of the dendrites undergo sideways migration and become enwrapped ("e") by OSCs, but a few remain unwrapped upon maturity (one being labeled by 'u' in black). Partially-wrapped ORN dendrites (one labeled by "p") may represent intermediate stages from unwrapped immature to enwrapped mature status. The apical junctional complex (yellow) can be distinguished into those homotypic cell–cell junctions between OSCs, heterotypic junctions between OSCs and ORN dendrites, and autotypic junctions that link neighboring pleats or folds of the same OSC, and form the mesenteries of enwrapped dendrites (mesodendrites, arrowheads). Modified with permits from [11].

In summary, based on their cytoarchitectonic relations with OSCs, ORN dendrites at the OE superficial layer could be subtyped as follows: i) The unwrapped dendrites are positioned at the inter-sustentacular cell borderlines, ii) The enwrapped dendrites are each enclosed in a vertical passage barrel within a single OSC. The plasma membrane of the enwrapped dendrite is closely apposed to, and linked by, intercellular junctions with host OSC plasma membrane at the dendritic passage barrel. The mesentery of an enwrapped dendrite (mesodendrite) tethers the dendritic passage barrel to the side of the host OSC, and is formed by closely apposed, cell junction-connected plasma membrane layers of neighboring pleats/thick folds of the host OSC. Up to six enwrapped dendrites could be seen in the confine of a single OSC apical process. A possible mechanism for the formation of the mesodendrite will be discussed below, iii) A partially wrapped dendrite is almost, but not yet completely, wrapped by an OSC. As discussed below, the partially wrapped dendrites probably represent the intermediate stage of ORN dendrites progressing from unwrapped to enwrapped status.

The above-observed cytoarchitectonic relations between ORN dendrites and OSCs confirm previous findings in the mouse, goldfish, rat, or man [3,9,27,28]. It then came, quite unexpectedly, to discover that a majority (~54%) of the ORN dendrites at the rat OE superficial layer are actually enwrapped by OSCs. Unwrapped ORN dendrites account for only ~28%, and partially-wrapped dendrites were even fewer (~18%) [11]. These quantitative data seemingly contradict the conventional belief of a "normal" unwrapped positioning for an absolute majority of ORN dendrites. The discrepancy might be attributed to the fact that few quantitative analyses were attempted previously.

The earlier notion seemed mainly based on qualitative observations by using high-power electron microscopy [3,9,27,28].

Cell junctions in the OE have been extensively studied, by using transmission or scanning electron microscopy, confocal microscopy, immunohistochemistry, molecular biology, and other approaches. The OE possesses various junctional structures between cells or cell parts of a same cell. Typical apical junctional belt of zonula occludens tight junctions and zonula adherens junctions is present at the OE apical layer [3,9,11,35,36]. There are puncta adherentia and desmosomes between the basolateral sides of OE cells [9,37]. Of particular interest, the adherens junctions between the ORNs and OSCs appear different from those between OSCs or between sustentacular and microvillar cells [9]. Gap junctions have also been reported in the OE, but molecular details and cellular distribution of gap junctions in the ORNs and OSCs remain unclear [38,39].

4. The Enwrapment, Sideways Migration and Terminal Maturation of ORN Dendrites

Class-III β-tubulin (Tuj1 immunoreactivity) is a known marker of immature ORNs. Its expression largely stops in ORN cell bodies and greatly weakens in ORN dendrites when ORNs start to show positivity for olfactory marker protein (a marker of mature ORNs) [9,40]. Thus, weak Tuj1 immunoreactivity indicates maturity of ORNs. In accordance with this notion, it was observed in the rat OE that intensely Tuj1-immunoreactive ORN cell bodies were mostly located near the basement membrane, whereas very weakly Tuj1-positive ORN cell bodies were mostly located in the OE middle layer [11].

ORN dendrites at the OE superficial layer also display significantly variable Tuj1 immunoreactivity intensities. Surprisingly, correlation of Tuj1 immunoreactivity intensities with wrapping status of ORN dendrites revealed that essentially all enwrapped dendrites have weak Tuj1 immunoreactivity. Strongly Tuj1-immunoreactive ORN dendrites are mostly located at the inter-sustentacular cell borderlines, and thus belong to the unwrapped subtype. A small number of the partially wrapped dendrites also exhibit high Tuj1 immunoreactivity. In view of the abovementioned reports of high Tuj1 immunoreactivity marking immature ORNs and weak Tuj1 immunoreactivity marking olfactory marker protein-expressing ORNs, these results indicate that the sustentacular cell-enwrapped dendrites mostly, if not all, belong to mature ORNs. The unwrapped and partially wrapped subtypes include practically all of the highly Tuj1-immunoreactive immature ORN dendrites and a portion of the low Tuj1-immunoreactive mature ORN dendrites. Quantitatively, the highly Tuj1-immunoreactive immature dendrites account for ~13% (10% unwrapped and 3% partially-wrapped) of all ORN dendrites at the OE apical surface, and the low-Tuj1-immunoreactive mature dendrites make up ~87% (54% enwrapped, 18% unwrapped, and 15% partially-wrapped). Overall, it appears that immature dendrites of newly generated ORNs mostly or all emerge from the borderlines between OSCs (unwrapped). This notion is consistent with previous observations by Nomura and coworkers [28] stating that "the dendrite of mature neurons was often wrapped by the supporting cells, while that of immature neurons was usually rather independent from the supporting cells".

The question then arises as to how the unwrapped immature ORN dendrites further mature after arriving at the OE apical surface, and eventually become mostly enwrapped [11]. Based on the presence of a mesentery (mesodendrite) trailing each enwrapped dendrite, the enwrapment appears to involve a sideways migration process of the newly-emerged immature dendrites from the inter-sustentacular borderlines to the intra-sustentacular enwrapped positions. The diversion point of the mesodendrite from the inter-sustentacular borderline was most likely the starting point, whereas the trajectory of the mesodendrite probably represents the path of the sideways migration process [11]. All enwrapped dendrites seem derived from the unwrapped. In other words, probably no ORN dendrite is enwrapped and mature upon first arrival at the OE luminal surface. This notion is supported by previous data indicating a loss of strong Tuj1 immunoreactivity in ORN dendrites only after the dendrites have inserted into the OE surface and started to produce olfactory marker protein [9].

Thus, the sideways migration appears a terminal maturation process undertaken by all enwrapped ORN dendrites. From this point of view, some or most of the partially wrapped dendrites possibly represent those at the intermediate stage of sideways migration from immature unwrapped to mature enwrapped status. Judging from the proportions of mature and immature ORN dendrites in the partially wrapped group (15% and 3%, respectively), it seems that most of the dendrites destined for OSC enwrapment could complete the terminal maturation process before reaching the final enwrapped position.

It should be noted that some 18% ORN dendrites remain unwrapped after maturity, as judged by their low levels of Tuj1 immunoreactivity. Currently, it is unknown whether these dendrites have also to undergo the process of sideways migration (understandably along the direction of inter-sustentacular borderlines, if at all) for terminal maturation at the OE luminal surface, or these dendrites mature in situ at the sites of emerging to the OE surface. In any case, the presence of a fraction of mature ORN dendrites at the inter-sustentacular borderlines suggests that OSC enwrapment is not really indispensable for terminal maturation and essential functionality of ORN dendrites. It awaits future investigations to elucidate the exact functional roles of ORN dendritic enwrapment. Among other possibilities, the enwrapment could help evenly disperse and insulate ORN dendrites and dendritic knobs on the OE surface to enhance olfactory reception or discrimination, or organize the mature dendrites into subgroups and modular units of olfactory reception, transduction, sensitivity, receptor potential transmission or other attributes. OSCs, for example, are reactive to purinergic and cholinergic neurochemicals, produce endocannabinoids, generate long-lasting positive potentials or propagatable Ca^{2+} signals when stimulated, and therefore could potentially modulate activities of ORN dendrites within its confine [10,14,16,41–44].

Thus ORN maturation perceivably involves not only the conventionally known vertical migration-and-sprouting phase for the cell bodies and dendrites, but also the newly discerned sideways migration terminal maturation phase for the enwrapped dendrites at least. The precision and complexity of the former phase have been fully appreciated [12,13,45–47], whereas the latter phase still awaits more in-depth investigations and better understanding. Given an estimated 40-day average lifespan of mature ORNs [18–21], the average immature-to-mature turnover time for the sideways migration and terminal maturation of newly emerged immature ORN dendrites is estimated to be about 5.98 days, based on the 13:87 ratio for percentage of immature-to-mature ORN dendrites at the OE luminal surface. In other words, after arriving at the OE surface, immature ORN dendrites spend another 5–6 days to further mature and downregulate own Tuj1 immunoreactivity. In reality, the turnover time should be even shorter as the turnover rate from immature to mature dendrites is unlikely to be always 100%.

5. Comparison of ORN Dendritic Enwrapment, Neurite Myelination, Remak's Bundling, and Possible Molecular Mechanisms

As far as the enwrapment of ORN dendrites is concerned, the OSCs somehow share similarities with the Schwann cells and oligodendrocytes of the peripheral and central nervous systems, respectively. The enwrapped ORN dendrites are somehow reminiscent of the myelinated nerve fibers, whereas the unwrapped ORN dendrites are reminiscent of the unmyelinated fibers. This poses the question as to whether the olfaction reception, signal transduction, or receptor potential propagation of enwrapped mature ORN dendrites may differ from that of the unwrapped mature ORN dendrites. Action potential conductions, for example, greatly differ between myelinated and unmyelinated nerve fibers [48,49]. At present, it is equally unclear if there is systematic variation in some other structural and functional properties between enwrapped and unwrapped ORN dendrites.

Cytologically, the relation between the enwrapped ORN dendrites and host OSCs might more resemble that between the unmyelinated C fibers and Schwann cells in the Remak's bundles of peripheral nerves, especially the sensory group-C fibers of the peripheral nervous system. If we consider direction of action or receptor potential propagation, the sensory C fibers in the peripheral nerves can indeed be viewed as dendrites or axodendrites of dorsal root ganglion neurons, much like the ORN dendrites. The so called mesaxon (mesentery of axon) associated with each of the C fibers in the Remak's bundle also bears great resemblance to the mesenteries of enwrapped ORN dendrites (mesodendrites). Remak's bundling of C fibers has possible functional roles in nerve development, differentiation or regeneration, and abnormalities of the Remak's bundles have been implicated in neurological disorders like neuropathies [49,50]. It remains to be clarified if ORN dendrite enwrapment may have similar functional roles in the development, differentiation, or regeneration of ORNs, and if abnormalities of ORN dendrite enwrapment may result in olfaction dysfunctions such as anosmia, parosmia, or phantosmia. This is especially worth noting in relation to well-documented olfaction abnormalities in aging, neurodegenerative, or neuropsychiatric disorders like Alzheimer's disease and schizophrenia [51–58].

At the molecular level, the ORNs share the common features of other neuronal cells of the nervous system, but the OSCs have been considered more as epithelial cells (as indicated by the cell type's expression of such marker proteins as keratin 8, E-Cadherin and Keratin 18 [59]). Few molecular similarities with neuroglia (such as Schwann cells or oligodendroglia) have been reported of the OSCs, with the exception of the actin cytoskeleton-related ezrin-radixin-moesin (ERM) protein family that serve as well-known linker molecules between cellular plasma membrane and actin cytoskeleton. Both the actin cytoskeleton and the ERM proteins play important roles not only in various epithelial cells, but also in the Schwann cells and oligodendroglia (in an event of myelination and/or formation of the node of Ranvier [60–64]). The ERMs are also abundantly expressed in the OSCs, especially at the apices and microvilli of OSCs [65].

More interestingly, juxtanodin (ermin), a protein molecule that shares the C-terminal actin-binding motif of the ERMs but lacks the N-terminal FERM (4.1 protein-ERM) domain thereof, was originally reported a specific oligodendroglial myelinic protein and a regulator of oligodendroglial actin cytoskeleton dynamics [66–69]. Specific juxtanodin (ermin) expression has subsequently been revealed in the OSCs and in retinal pigment epithelial cells [17,70]. These data add to the molecular commonalties among the myelin-forming and special sense organ supporting cells, and strongly implicate the actin cytoskeleton and related proteins in regulating ORN dendrite wrapping by OSCs. These commonalities also suggest shared molecular mechanisms between myelination and enwrapment of ORN dendrites. The retinal photoreceptor neurons (rods and cones) notably are also partly surrounded by retinal pigment epithelial cells. However, rather than the sideways invagination of ORN dendrites into OSCs, the outer segments of retinal rods and cones approach the retinal pigment epithelial cells from the opposite direction, and interdigitate with the microvilli and protrusions of the latter [71].

6. Some Other Questions and Future Studies

The finding of ORN dendrite enwrapment and sideways migration updated our understanding of the OE cytoarchitecture. It also raised many new questions that should be addressed in future studies. First, in terms of phylogeny and ontogeny, quantitative data of ORN dendrite enwrapment and lateral migration maturation is not available for comparison among species, organisms, or developmental and aging stages of individual species. Qualitatively, it has been reported that the mouse has relatively more sustentacular cell-enclosed ORN dendrites than the goldfish, and sustentacular cell-enclosed ORN dendrites were seen not only in the adult mouse OE, but also in the fetus and newborn mice [3]. Concerning the latter, it should be noted that ORN dendrites in the fetus or newborn may not necessarily be immature, as olfactory marker protein-positive ORNs and dendrites are already present in the embryo [27]. Further investigations of the enwrapment and sideways migration of ORN dendrites in different species and developmental or aging stages would help understand the biology and functions of the OE.

Second, detailed cytoarchitectonic relations among the cell types at middle and basal layers of the OE, especially between OSC foot processes and ORN cell bodies and axons, remain unclear. Limited electron microscopic studies have suggested the presence of partially wrapped ORN cell bodies and axons [27,28]. Further investigation of the issue would elucidate if sustentacular wrapping/enclosure, complete or partial, has a role in development, differentiation or vertical migration guidance of immature ORNs, or in the elongation and selective bundling of ORN axons.

More importantly, are there possible molecular or functional differences among the enwrapped, partially-wrapped and unwrapped mature ORN dendrites? Among others, possible differences in odorant receptor expression profiles remain elusive between the mature enwrapped and mature unwrapped ORN dendrites, among ORN dendrites wrapped within individual OSCs, or between ORN dendrites enwrapped by different OSCs. Moreover, are there variations in ORN dendrite enwrapment across different zones/regions of the OE in relation to the systematic variations of odorant receptor expression across the zones [12,13], or in relation to the different subsystems of odorant receptors like the class-I and class-II canonical odorant receptors or trace amine-associated receptors? [72–76] Further clarifications on these and related other issues would not only enlighten us on the biological and functional meaning of ORN dendrite enwrapment and differentiation, but may also implicate the wrapping status of ORN dendrites in organizing olfactory subsystems.

Finally, what are the possible pathological changes of the ORN dendrite wrapping status and sideways migration in anosmia, parosmia, neurodegenerative, and neuropsychiatric disorders? Olfaction dysfunctions are frequent early manifestations of neuropsychiatric and aging-related neurodegenerative disorders like schizophrenia and Alzheimer's disease [51–58]. It is thus particularly relevant to investigate possible OE cytoarchitectonic and biochemical alterations in patients inflicted with brain aging and mental disorders. Unlike the CNS, the OE is accessible for endoscopy and biopsy examinations, and OE mucus/swabs can be readily obtained and tested for cytological or biochemical abnormalities.

Funding: The author's research is supported by grants from the Academic Research Funds (AcRF) (Ministry of Health, Singapore, R-181-000-182-114), the Singapore Biomedical Research Council (BMRC/04/1/21/19/305), and the National Medical Research Council (Singapore) (0946/2005).

Conflicts of Interest: The authors declare no conflicts of interest. The funders had no role in the design of the study; in the collection, analyses, or interpretation of data; in the writing of the manuscript; or in the decision to publish the results.

References

1. Schultze, M. Über die endigungsweise des geruchsnerven und die epithelialgebilde der nasenschleimhaut. *Monatsberichte der Konigl Preußs. Akad der Wissen Berlin* **1856**, *21*, 504–515.
2. Allison, A.C. The morphology of the olfactory system in vertebrates. *Biol. Rev.* **1953**, *28*, 195–244. [CrossRef]
3. Breipohl, W.; Laugwitz, H.J.; Bornfeld, N. Topological relations between the dendrites of olfactory sensory cells and sustentacular cells in different vertebrates. An ultrastructural study. *J. Anat.* **1974**, *117*, 89–94. [PubMed]
4. Suzuki, Y.; Takeda, M.; Farbman, A.I. Supporting cells as phagocytes in the olfactory epithelium after bulbectomy. *J. Comp. Neurol.* **1996**, *376*, 509–517. [CrossRef]
5. Schwob, J.E. Neural regeneration and the peripheral olfactory system. *Anat. Rec.* **2002**, *269*, 33–49. [CrossRef] [PubMed]
6. Menco, B.P.H.M.; Morrison, E.E. Morphology of the mammalian olfactory epithelium: Form, fine structure and pathology. In *Handbook of Olfaction and Gustation*, 2nd ed.; Doty, R., Ed.; Marcel Dekker: New York, NY, USA, 2003; pp. 17–49.
7. Asan, E.; Drenckhahn, D. Immunocytochemical characterization of two types of microvillar cells in rodent olfactory epithelium. *Histochem. Cell Biol.* **2005**, *123*, 157–168. [CrossRef]
8. Elsaesser, R.; Paysan, J. The sense of smell, its signalling pathways, and the dichotomy of cilia and microvilli in olfactory sensory cells. *BMC Neurosci.* **2007**, *8*, S1. [CrossRef]
9. Steinke, A.; Meier-Stiegen, S.; Drenckhahn, D.; Asan, E. Molecular composition of tight and adherens junctions in the rat olfactory epithelium and fila. *Histochem. Cell Biol.* **2008**, *130*, 339. [CrossRef]
10. Hegg, C.C.; Irwin, M.; Lucero, M.T. Calcium store-mediated signaling in sustentacular cells of the mouse olfactory epithelium. *Glia* **2009**, *57*, 634–644. [CrossRef]

11. Liang, F. Olfactory receptor neuronal dendrites become mostly intra-sustentacularly enwrapped upon maturity. *J. Anat.* **2018**, *232*, 674–685. [CrossRef]
12. Buck, L.B. Unraveling the sense of smell (Nobel lecture). *Angew. Chem. Int. Ed. Engl.* **2005**, *44*, 6128–6140. [CrossRef] [PubMed]
13. Mori, K.; von Campenhause, H.; Yoshihara, Y. Zonal organization of the mammalian main and accessory olfactory systems. *Philos. Trans. R. Soc. Lond. B Biol. Sci.* **2000**, *355*, 1801–1812. [CrossRef] [PubMed]
14. Okano, M.; Takagi, S.F. Secretion and electrogenesis of the supporting cell in the olfactory epithelium. *J. Physiol.* **1974**, *242*, 353–370. [CrossRef] [PubMed]
15. Suzuki, Y.; Takeda, M.; Obara, N.; Suzuki, N.; Takeichi, N. Olfactory epithelium consisting of supporting cells and horizontal basal cells in the posterior nasal cavity of mice. *Cell Tissue Res.* **2000**, *299*, 313–325. [CrossRef]
16. Hassenklöver, T.; Kurtanska, S.; Bartoszek, I.; Junek, S.; Schild, D.; Manzini, I. Nucleotide-induced Ca^{2+} signaling in sustentacular supporting cells of the olfactory epithelium. *Glia* **2008**, *56*, 1614–1624. [CrossRef] [PubMed]
17. Tang, J.; Tang, J.H.; Ling, E.A.; Wu, Y.; Liang, F. Juxtanodin in the rat olfactory epithelium: Specific expression in sustentacular cells and preferential subcellular positioning at the apical junctional belt. *Neuroscience* **2009**, *161*, 249–258. [CrossRef]
18. Graziadei, P.P.C.; Graziadei, G.A.M. Neurogenesis and neuron regeneration in the olfactory system of mammals. I. Morphological aspects of differentiation and structural organization of the olfactory sensory neurons. *J. Neurocytol.* **1979**, *8*, 1–18. [CrossRef] [PubMed]
19. Shipley, M.T.; Ennis, M.; Puche, A.C. Olfactory system. In *The Rat Nervous System*, 3rd ed.; Paxinos, G., Ed.; Elsevier: San Diego, CA, USA, 2004; pp. 923–964.
20. Ashwell, K. The olfactory system. In *The Mouse Nervous System*; Watson, C., Paxinos, G., Puelles, L., Eds.; Academic Press: Amsterdam, The Netherland, 2012; pp. 653–660.
21. Brann, J.H.; Firestein, S.J. A lifetime of neurogenesis in the olfactory system. *Front. Neurosci.* **2014**, *8*, 182. [CrossRef]
22. Kwon, B.S.; Kim, M.K.; Kim, W.H.; Pyo, J.S.; Cheon, Y.H.; Cha, C.I.; Nam, S.Y.; Baik, T.K.; Lee, B.L. Age-related changes in microvillar cells of rat olfactory epithelium. *Neurosci. Lett.* **2005**, *378*, 65–69. [CrossRef]
23. Montani, G.; Tonelli, S.; Elsaesser, R.; Paysan, J.; Tirindelli, R. Neuropeptide Y in the olfactory microvillar cells. *Eur. J. Neurosci.* **2006**, *24*, 20–24. [CrossRef] [PubMed]
24. Lemons, K.; Fu, Z.; Aoudé, I.; Ogura, T.; Sun, J.; Chang, J.; Mbonu, K.; Matsumoto, I.; Arakawa, H.; Lin, W. Lack of TRPM5-expressing microvillous cells in mouse main olfactory epithelium leads to impaired odor-evoked responses and olfactory-guided behavior in a challenging chemical environment. *eNeuro* **2017**, *4*, ENEURO.0135-17.2017. [CrossRef] [PubMed]
25. Genovese, F.; Tizzano, M. Microvillous cells in the olfactory epithelium express elements of the solitary chemosensory cell transduction signaling cascade. *PLoS ONE* **2018**, *13*, e0202754. [CrossRef] [PubMed]
26. Graziadei, P.P.C. The olfactory mucosa of vertebrates. In *Olfaction. Handbook of Sensory Physiology*; Beidler, L.M., Ed.; Springer: Berlin, Germany, 1971; Volume 4/1.
27. Morrison, E.E.; Costanzo, R.M. Morphology of olfactory epithlium in humans and other vertebrates. *Microsc. Res. Tech.* **1992**, *23*, 49–61. [CrossRef] [PubMed]
28. Nomura, T.; Takahashi, S.; Ushiki, T. Cytoarchitecture of the normal rat olfactory epithelium: Light and scanning electron microscopic studies. *Arch. Histol. Cytol.* **2004**, *67*, 159–170. [CrossRef]
29. Ma, M.; Shepherd, G.M. Functional mosaic organization of mouse olfactory receptor neurons. *Proc. Natl. Acad. Sci. USA* **2000**, *97*, 12869–12874. [CrossRef]
30. Barral, J.-P.; Croibier, A. Chapter 10: Olfactory nerve. In *Manual Therapy for the Cranial Nerves*; Churchill Livingstone: London, UK, 2009; p. 61. ISBN 9780702031007.
31. Farbman, A.I. *Cell Biology of Olfaction*; Cambridge University Press: New York, UK, USA, 1992.
32. Smutzer, G.S.; Doty, R.L.; Arnold, S.E.; Trojanowski, J.Q. Olfactory system neuropathology in Alzheimer's disease, Parkinson's disease, and schizophrenia. In *Handbook of Olfaction and Gustation*, 2nd ed.; Doty, R., Ed.; Marcel Dekker: New York, USA, 2003; pp. 503–523.
33. Standring, S. *Gray's Anatomy: The Anatomical Basis of Clinical Practice*, 40th ed.; Churchill-Livingstone: London, UK; Elsevier: Amsterdam, The Netherlands, 2008; pp. 552–553.
34. Salazar, I.; Sanchez-Quinteiro, P.; Barrios, A.W.; López Amado, M.; Vega, J.A. Anatomy of the olfactory mucosa. *Handb. Clin. Neurol.* **2019**, *164*, 47–65. [CrossRef]

35. Menco, B.P.H.M. Qualitative and quantitative freeze-fracture studies on olfactory and nasal respiratory epithelial surfaces of frog, ox, rat, and dog III. Tight-junctions. *Cell Tissue Res.* **1980**, *211*, 361–373. [CrossRef]
36. Wolburg, H.; Wolburg-Buchholz, K.; Sam, H.; Horvát, S.; Deli, M.A.; Mack, A.F. Epithelial and endothelial barriers in the olfactory region of the nasal cavity of the rat. *Histochem. Cell Biol.* **2008**, *130*, 127–140. [CrossRef]
37. Moran, D.T.; Rowley, J.C., 3rd; Jafek, B.W.; Lovell, M.A. The fine structure of the olfactory mucosa in man. *J. Neurocytol.* **1982**, *11*, 721–746. [CrossRef]
38. Vogalis, F.; Hegg, C.C.; Lucero, M.T. Electrical coupling in sustentacular cells of the mouse olfactory epithelium. *J. Neurophysiol.* **2005**, *94*, 1001–1012. [CrossRef]
39. Zhang, C.B. Gap junctions in olfactory neurons modulate olfactory sensitivity. *BMC Neurosci.* **2010**, *11*, 108. [CrossRef] [PubMed]
40. Roskams, A.J.; Cai, X.; Ronnet, G.V. Expression of neuron-specific beta-III tubulin during olfactory neurogenesis in the embryonic and adult rat. *Neuroscience* **1998**, *83*, 191–200. [CrossRef]
41. Czesnik, D.; Schild, D.; Kuduz, J.; Manzini, I. Cannabinoid action in the olfactory epithelium. *Proc. Natl. Acad. Sci. USA* **2007**, *104*, 2967–2972. [CrossRef] [PubMed]
42. Breunig, E.; Manzini, I.; Piscitelli, F.; Gutermann, B.; Di Marzo, V.; Schild, D.; Czesnik, D. The endocannabinoid 2-arachidonoyl-glycerol controls odor sensitivity in larvae of *Xenopus laevis*. *J. Neurosci.* **2010**, *30*, 8965–8973. [CrossRef]
43. Ogura, T.; Szebenyi, S.A.; Krosnowski, K.; Sathyanesan, A.; Jackson, J.; Lin, W. Cholinergic microvillous cells in the mouse main olfactory epithelium and effect of acetylcholine on olfactory sensory neurons and supporting cells. *J. Neurophysiol.* **2011**, *106*, 1274–1287. [CrossRef]
44. Hutch, C.R.; Hegg, C.C. Cannabinoid receptor signaling induces proliferation but not neurogenesis in the mouse olfactory epithelium. *Neurogenesis* **2016**, *3*, e1118177. [CrossRef]
45. Fletcher, R.B.; Das, D.; Gadye, L.; Street, K.N.; Baudhuin, A.; Wagner, A.; Cole, M.B.; Flores, Q.; Choi, Y.G.; Yosef, N.; et al. Deconstructing olfactory stem cell trajectories at single-cell resolution. *Cell Stem Cell* **2017**, *20*, 817–830. [CrossRef]
46. Coleman, J.H.; Lin, B.; Louie, J.D.; Peterson, J.; Lane, R.P.; Schwob, J.E. Spatial determination of neuronal diversification in the olfactory epithelium. *J. Neurosci.* **2019**, *39*, 814–832. [CrossRef]
47. Li, H.; Li, T.; Horns, F.; Li, J.; Xie, Q.; Xu, C.; Wu, B.; Kebschull, J.M.; McLaughlin, C.N.; Kolluru, S.S.; et al. Single-cell transcriptomes reveal diverse regulatory strategies for olfactory receptor expression and axon targeting. *Curr. Biol.* **2020**, *30*, 1189–1198. [CrossRef]
48. Craig, A.D. How do you feel? Interoception: The sense of the physiological condition of the body. *Nat. Rev. Neurosci.* **2002**, *3*, 655–666. [CrossRef]
49. Murinson, B.B.; Griffin, J.W. C-fiber structure varies with location in peripheral nerve. *J. Neuropathol. Exp. Neurol.* **2004**, *63*, 246–254. [CrossRef] [PubMed]
50. Harty, B.L.; Monk, K.R. Unwrapping the unappreciated: Recent progress in Remak Schwann cell biology. *Curr. Opin. Neurobiol.* **2017**, *47*, 131–137. [CrossRef] [PubMed]
51. Arnold, S.E.; Smutzer, G.S.; Trojanowski, J.Q.; Moberg, P.J. Cellular and molecular neuropathology of the olfactory epithelium and central olfactory pathways in Alzheimer's disease and schizophrenia. *Ann. N. Y. Acad. Sci.* **1998**, *855*, 762–775. [CrossRef] [PubMed]
52. Doty, R.L. Olfaction in Parkinson's disease and related disorders. *Neurobiol. Dis.* **2012**, *46*, 527–552. [CrossRef] [PubMed]
53. Buron, E.; Bulbena, A. Olfaction in affective and anxiety disorders: A review of the literature. *Psychopathology* **2013**, *46*, 63–74. [CrossRef] [PubMed]
54. Casjens, S.; Eckert, A.; Woitalla, D.; Ellrichmann, G.; Turewicz, M.; Stephan, C.; Eisenacher, M.; May, C.; Meyer, H.E.; Brüning, T.; et al. Diagnostic value of the impairment of olfaction in Parkinson's disease. *PLoS ONE* **2013**, *8*, e64735. [CrossRef]
55. Seligman, S.C.; Kamath, V.; Giovannetti, T.; Arnold, S.E.; Moberg, P.J. Olfaction and apathy in Alzheimer's disease, mild cognitive impairment, and healthy older adults. *Aging Ment. Health* **2013**, *17*, 564–570. [CrossRef]
56. Auster, T.L.; Cohen, A.S.; Callaway, D.A.; Brown, L.A. Objective and subjective olfaction across the schizophrenia spectrum. *Psychiatry* **2014**, *77*, 57–66. [CrossRef]

57. Doty, R.L.; Hawkes, C.H.; Good, K.P.; Duda, J.E. Odor perception and neuropathology in neurodegenerative diseases and schizophrenia. In *Handbook of Olfaction and Gustation*; Doty, R.L., Ed.; Wiley: Hoboken, NJ, USA, 2015; pp. 93–108.
58. Field, T. Smell and taste dysfunction as early markers for neurodegenerative and neuropsychiatric diseases. *J. Alzheimers Dis. Parkinsonism* **2015**, *5*, 186.
59. Holbrook, E.H.; Wu, E.; Curry, W.T.; Lin, D.T.; Schwob, J.E. Immunohistochemical characterization of human olfactory tissue. *Laryngoscope* **2011**, *121*, 1687–1701. [CrossRef] [PubMed]
60. Melendez-Vasquez, C.V.; Rios, J.C.; Zanazzi, G.; Lambert, S.; Bretscher, A.; Salzer, J.L. Nodes of Ranvier form in association with ezrin-radixin-moesin (ERM)-positive Schwann cell processes. *Proc. Natl. Acad. Sci. USA* **2001**, *98*, 1235–1240. [CrossRef]
61. Gatto, C.L.; Walker, B.J.; Lambert, S. Local ERM activation and dynamic growth cones at Schwann cell tips implicated in efficient formation of nodes of Ranvier. *J. Cell Biol.* **2003**, *162*, 489–498. [CrossRef]
62. Nawaz, S.; Sánchez, P.; Schmitt, S.; Snaidero, N.; Mitkovski, M.; Velte, C.; Brückner, B.R.; Alexopoulos, I.; Czopka, T.; Jung, S.Y.; et al. Actin filament turnover drives leading edge growth during myelin sheath formation in the central nervous system. *Dev. Cell* **2015**, *34*, 139–151. [CrossRef] [PubMed]
63. Samanta, J.; Salzer, J.L. Myelination: Actin disassembly leads the way. *Dev. Cell* **2015**, *34*, 129–130. [CrossRef] [PubMed]
64. Zuchero, J.B.; Fu, M.M.; Sloan, S.A.; Ibrahim, A.; Olson, A.; Zaremba, A.; Dugas, J.C.; Wienbar, S.; Caprariello, A.V.; Kantor, C.; et al. CNS myelin wrapping is driven by actin disassembly. *Dev. Cell* **2015**, *34*, 152–167. [CrossRef] [PubMed]
65. Maurya, D.K.; Henriques, T.; Marini, M.; Pedemonte, N.; Galietta, L.J.; Rock, J.R.; Harfe, B.D.; Menini, A. development of the olfactory epithelium and nasal glands in TMEM16A$^{-/-}$ and TMEM16A$^{+/+}$ mice. *PLoS ONE* **2015**, *10*, e0129171. [CrossRef]
66. Zhang, B.; Cao, Q.; Guo, A.; Chu, H.; Chan, Y.G.; Buschdorf, J.P.; Low, B.C.; Ling, E.A.; Liang, F. Juxtanodin: An oligodendroglial protein that promotes cellular arborization and 2′,3′-cyclic nucleotide-3′-phosphodiesterase trafficking. *Proc. Natl. Acad. Sci. USA* **2005**, *102*, 11527–11532. [CrossRef]
67. Brockschnieder, D.; Sabanay, H.; Riethmacher, D.; Peles, E. Ermin, a myelinating oligodendrocyte-specific protein that regulates cell morphology. *J. Neurosci.* **2006**, *26*, 757–762. [CrossRef]
68. Meng, J.; Xia, W.; Tang, J.H.; Tang, B.L.; Liang, F. Dephosphorylation-dependent inhibitory activity of juxtanodin on filamentous actin disassembly. *J. Biol. Chem.* **2010**, *285*, 28838–28849. [CrossRef]
69. Ruskamo, S.; Chukhlieb, M.; Vahokoski, J.; Bhargav, S.P.; Liang, F.; Kursula, I.; Kursula, P. Juxtanodin is an intrinsically disordered F-actin-binding protein. *Sci. Rep.* **2012**, *2*, 899. [CrossRef]
70. Liang, F.; Hwang, J.H.; Tang, N.W.; Hunziker, W. Juxtanodin in retinal pigment epithelial cells: Expression and biological activities in regulating cell morphology and actin cytoskeleton organization. *J. Comp. Neurol.* **2018**, *526*, 205–215. [CrossRef] [PubMed]
71. Matsumoto, B.; Defoe, D.M.; Besharse, J.C. Membrane turnover in rod photoreceptors: Ensheathment and phagocytosis of outer segment distal tips by pseudopodia of the retinal pigment epithelium. *Proc. R. Soc. Lond. B Biol. Sci.* **1987**, *230*, 339–354. [PubMed]
72. Buck, L.B.; Axel, R. A novel multigene family may encode odorant receptors: A molecular basis for odor recognition. *Cell* **1991**, *65*, 157–167. [CrossRef]
73. Fleischer, J.; Breer, H.; Strotmann, J. Mammalian olfactory receptors. *Front. Cell. Neurosci.* **2009**, *3*, 9. [CrossRef]
74. Johnson, M.A.; Tsai, L.; Roy, D.S.; Valenzuela, D.H.; Mosley, C.; Magklara, A.; Lomvardas, S.; Liberles, S.D.; Barnea, G. Neurons expressing trace amine-associated receptors project to discrete glomeruli and constitute an olfactory subsystem. *Proc. Natl. Acad. Sci. USA* **2012**, *109*, 13410–13415. [CrossRef]
75. Ihara, S.; Yoshikawa, K.; Touhara, K. Chemosensory signals and their receptors in the olfactory neural system. *Neuroscience* **2013**, *254*, 45–60. [CrossRef]
76. Bear, D.M.; Lassance, J.M.; Hoekstra, H.E.; Datta, S.R. The evolving neural and genetic architecture of vertebrate olfaction. *Curr. Biol.* **2016**, *26*, R1039–R1049. [CrossRef]

Review

Cannabinoid Control of Olfactory Processes: The *Where* Matters

Geoffrey Terral [1,2,3], Giovanni Marsicano [1,2], Pedro Grandes [4,5] and Edgar Soria-Gómez [4,5,6,*]

1 INSERM, U1215 NeuroCentre Magendie, 146 rue Léo Saignat, CEDEX, 33077 Bordeaux, France; geoffreyterral@gmail.com (G.T.); giovanni.marsicano@inserm.fr (G.M.)
2 University of Bordeaux, 146 rue Léo Saignat, 33000 Bordeaux, France
3 Interdisciplinary Institute for Neuroscience, CNRS, UMR 5297, 33000 Bordeaux, France
4 Department of Neurosciences, University of the Basque Country UPV/EHU, Barrio Sarriena s\n, 48940 Leioa, Spain; pedro.grandes@ehu.eus
5 Achucarro Basque Center for Neuroscience, Science Park of the UPV/EHU, 48940 Leioa, Spain
6 IKERBASQUE, Basque Foundation for Science, Maria Diaz de Haro 3, 48013 Bilbao, Spain
* Correspondence: edgarjesus.soria@ehu.eus or edgar.soria@achucarro.org

Received: 25 February 2020; Accepted: 13 April 2020; Published: 16 April 2020

Abstract: Olfaction has a direct influence on behavior and cognitive processes. There are different neuromodulatory systems in olfactory circuits that control the sensory information flowing through the rest of the brain. The presence of the cannabinoid type-1 (CB1) receptor, (the main cannabinoid receptor in the brain), has been shown for more than 20 years in different brain olfactory areas. However, only over the last decade have we started to know the specific cellular mechanisms that link cannabinoid signaling to olfactory processing and the control of behavior. In this review, we aim to summarize and discuss our current knowledge about the presence of CB1 receptors, and the function of the endocannabinoid system in the regulation of different olfactory brain circuits and related behaviors.

Keywords: olfaction; endocannabinoids; olfactory epithelium; olfactory bulb; piriform cortex; CB1 receptor

---

1. The Endocannabinoid System: A General Overview

Cannabis sativa, also known as marijuana or cannabis, has been used for thousands of years for its therapeutic and recreational properties. Nowadays, after tobacco and alcohol, cannabis is the most commonly consumed drug of abuse, with 188 million cannabis users estimated worldwide in 2017 [1]. The cannabinoid receptors type-1 (CB1) and type-2 (CB2), their endogenous ligands (endocannabinoids), and the synthetic and degradative enzymes that regulate endocannabinoid levels support the concept of the endocannabinoid system (ECS) as participating in the regulation of physiological processes [2]. CB1 and CB2 receptors belong to the superfamily of G protein-coupled receptors (GPCRs) that consist of seven transmembrane domains with an extracellular N-terminal and an intracellular C-terminal tail [3]. At the synaptic level, endocannabinoids can be synthesized, but not exclusively [4], by post-synaptic intracellular calcium elevations, which can be caused by various stimuli, including depolarization, the activation of metabotropic acetylcholine, and glutamate receptors, particularly Gq-coupled receptors (i.e., M1/M3 and mGluR 1/5) [3]. Once produced, endocannabinoids act on CB1 receptors that are mainly described at pre-synaptic terminals [2,3,5], and other cellular locations [6].

In neurons, the main effect of CB1 receptor activation is a decrease in neurotransmitter release, inducing different forms of endocannabinoid-mediated plasticity [3], such as the depolarization-induced suppression of inhibition/excitation (DSI/DSE; [7–9]), or the long-term depression of inhibitory/

excitatory synapses [10–14]. CB1 receptors are widely expressed in the central nervous system and likely represent the most abundant GPCR in the brain [15]. Given its ubiquitous expression in multiple brain areas, CB1 receptors modulate a variety of functions, from sensory perception to more complex cognitive processes such as learning and memory [16–18].

2. Role of the Endocannabinoid System in Olfactory Circuits

Known for a long time, one of the predominant subjective effects of cannabis intoxication is the alteration of sensory perception, including olfactory processes [19]. However, although relatively high levels of CB1 receptors were described in the 1990s in many olfactory brain areas of rodents [20–23], their olfactory-related functions only started to be studied during the last decade. Notably, the involvement of CB1 receptors in specific odor-related processes has been reported in specialized olfactory structures such as the olfactory epithelium (OE; [24–27]), the main olfactory bulb (MOB; [18,28–34]), and the piriform cortex (PC; [35–40]), but also in other brain areas processing olfactory information [41–44]. For the sake of clarity in this review, we will focus on describing the role of the ECS, particularly CB1 receptor signaling, in specific main olfactory areas (i.e., OE, MOB, and PC).

3. The Endocannabinoid System in the Olfactory Epithelium

The first hypothesis for the physiological involvement of endocannabinoids in olfactory processes came from three observations: (1) The olfactory perception was shown to be changed depending on the feeding state of individuals [45,46], (2) the ECS were proposed to be involved in food intake [18,47], and (3) the anatomical and functional connectivity between peripheral organs regulating energy balance and olfactory structures [18,48]. Czesnik, Breunig, and colleagues [24,25] provided the first evidence that cannabinoids could modulate olfaction. These studies revealed the presence of CB1 receptors in the olfactory sensory neurons (OSN) of Xenopus laevis and demonstrated that endocannabinoids modulate odor-evoked responses. Additionally, they found that the production of endocannabinoids depends on the hunger state of the animal, which is responsible for changes in odor sensitivity activity. Similarly, CB1 receptors were also found in the OSN of rodents [27]. The CB1 receptor agonists changed odorant-induced cellular activity, but the authors did not observe olfactory consequences in the behavior of mutant mice lacking CB1 receptors (CB1-KO; [27]). Despite the species differences, several divergences appear between these studies. For instance, the first two studies evaluated the impact of cannabinoids on odor sensitivity by recording the cellular activity of the OSN with calcium imaging and electrophysiological methods [24,25]. Instead, Hutch and colleagues [27] investigated the involvement of CB1 receptors in olfactory-mediated learning and memory tasks such as the buried food test and a habituation/dishabituation paradigm. In addition, CB1-KO mice lack brain specificity and might be confounded by compensatory mechanisms [49]. Thus, the physiological role of CB1 receptors in the mammalian OE still remains unclear and will need further investigation.

4. The Endocannabinoid System in the Olfactory Bulb

In the mammalian MOB, the ECS was first described as a modulator of GABAergic transmission [28,33]. Pharmacological approaches, combined with in vitro patch-clamp experiments, highlighted that CB1 receptors modulate the firing pattern of periglomerular (PG) and external tufted cells (eTCs). Considering that PG cells form synapses with mitral and tufted cells [50], CB1 signaling may indirectly regulate the main output activity of the MOB neurons. Indeed, the inhibitory inputs of eTCs display spontaneous DSI [33], and the pharmacological manipulation of CB1 receptor signaling modulates mitral cell activity, likely through indirect control of inhibitory transmission [34]. These results suggest that endocannabinoids are capable of controlling mitral/tufted cell activity through the CB1 receptors on PG cells. Although the authors did not investigate the behavioral impact of these effects, the CB1 receptors' activation may increase the signal-to-noise ratio and, thus, the overall sensitivity of the glomerulus to sensory inputs. Moreover, CB1 receptors are present in glutamatergic corticofugal fibers (CFF) coming from projection neurons from anterior cortical olfactory areas (including the anterior olfactory nucleus, AON, and the anterior piriform cortex), and targeting granule cells (GCs) of the MOB [32]. Consistent with the idea that cannabinoid signaling in the olfactory system might control the feeding state of the organism, the hypophagic phenotype observed in mice lacking CB1 receptors in their glutamatergic neurons is associated with an increased activity of CFF onto GCs. Notably, endocannabinoid levels increase in the MOB during fasting, allowing for the dampening of the excitation of GCs. Given that GCs control mitral cell activity, CB1 receptor activation of CFF induces the disinhibition of mitral cells. This effect is followed by a fasting-related enhancement in olfactory sensitivity, which correlates with the amount of food ingested upon refeeding. These results suggest that the endocannabinoid-mediated regulation of olfactory output information controls olfactory perception and food intake [32]. Since CB1 receptors have been described as being expressed on CFF fibers, they may thus regulate all of the downstream synapses of these fibers. This hypothesis was recently verified in the synapse between the CFF and the so-called deep short axon cells (dSAs; [31]). Indeed, depolarization of dSAs in the MOB of mice elicits pre-synaptic CB1 receptors' transient suppression of excitatory CFF inputs (DSE). In addition, the authors demonstrated that dSAs could inhibit GCs, thereby suppressing GC to mitral cells inhibition. Interestingly, depending on the CFF synaptic strength, the CB1 receptor signaling can either control the synapses from dSAs to GCs, or directly from GCs to mitral cells, suggesting a double dissociation in the control of olfactory bulb output neurons [31]. However, the behavioral consequences of this bidirectional effect remain to be elucidated.

5. The Endocannabinoid System in the Piriform Cortex

The PC is a brain area capable of generating epileptiform activity [51]. In other brain structures such as the hippocampus, CB1 receptors have been shown to protect against seizures [52,53]. Thus, the anticonvulsant effects of cannabinoids were assessed in PC slices [36]. The authors demonstrated that CB1 receptor agonists reduce seizures, indicating that CB1 receptor activation is able to control PC activity [36]. However, there is currently no functional evidence about how the ECS could affect olfactory processes under pathological conditions such as epilepsy. Furthermore, the ECS in the PC indirectly affects social behavior [38]. Although it does not affect social interactions per se, local injections of a CB1 receptor antagonist into the posterior PC (pPC) reversed the impairment of social sniffing time induced by an activation of dopamine receptors, suggesting that the ECS in the pPC has a deleterious effect on social behavior when coupled with dopamine activation [38]. Moreover, the PC is an important area involved in olfactory memory [54]. Considering that the ECS is highly studied in learning and memory functions [16], other studies investigated its role in PC-dependent olfactory learning and memory. In the pPC, odor-discrimination training leads to the endocannabinoid-mediated modification of inhibitory synapses [35]. Indeed, the learning of a complex olfactory rule induces the activation of CB1 receptors, which in turn enhances GABAergic conductance in post-synaptic pPC pyramidal neurons, indicating a postsynaptic effect [35]. Despite the possible post-synaptic CB1 receptors' localization, or that endocannabinoids can modulate directly postsynaptic GABAergic receptors [55], further experiments will determine how CB1 receptor activation allows for controlling GABAergic conductance in the pPC. In the anterior PC (aPC), CB1 receptors were mainly described at GABAergic synapses, where they modulate inhibitory transmission and plasticity [37,40]. Moreover, depending on CB1 receptors in the aPC, the retrieval of appetitive, but not aversive, olfactory memory is associated with a modulation of local inhibitory transmission onto specific principal cells in the aPC [37]. These data indicate that CB1 receptors in the aPC selectively control olfactory memory retrieval related to positively motivated behaviors. Thus, it will be crucial to determine if cannabinoid signaling controls functional connection between the PC and brain regions controlling affective states such as the orbitofrontal cortex, the nucleus accumbens or amygdala. In fact, there is compelling literature demonstrating the participation of these brain areas in olfactory processes [56–58]. In line with this idea, recent observations in humans highlight that the state-dependent enhancement of endocannabinoid levels changes dietary choices toward high-energy food items. Interestingly, this phenomenon was related to an increase in odor responses in the PC [39].

6. Conclusions

Growing evidence has revealed that the ECS modulates direct olfactory processes such as odor sensitivity or olfactory learning and memory. Across different brain olfactory areas, the ECS appears to play an essential role in the control of synaptic transmission and plasticity, but also in the regulation of vital behaviors that depend on olfaction, such as the feeding state of the individual (Figure 1; [59]). However, the physiological impact of the endocannabinoid-mediated plasticity, the contribution of CB1 receptors during other olfactory-dependent behaviors, and the contribution of each olfactory brain region (e.g., OE, MOB, PC) during specific behaviors remain to be elucidated. Furthermore, the role of other components of the ECS in olfactory processes is less clear, such as the role of CB2 receptors, which are described to be present in the OE [27]. This highlights the importance of continuing with this exciting line of research. To the same extent, there is a lack of direct evidence about the participation of other olfactory-related structures in cannabinoid-mediated effects, for example, the olfactory tubercle (OT). In fact, the OT is a target of different hormones and local modulators regulating feeding behavior and motivation [48]. Thus, it is reasonable to think about a potential cross-link between cannabinoid signaling and hormonal regulation in olfactory related behaviors taking place in the OT.

Besides the studies regarding the functions of the ECS in primary olfactory structures, it is important to take into account that CB1 receptors are present and modulate associated olfactory areas (i.e., amygdala, orbitofrontal cortex, hippocampus or periaqueductal gray; [41–44]), suggesting that olfactory processing that involves the control of different brain structures might also be modulated by the ECS. In humans, the main psychoactive compound of cannabis has been shown to induce an increase in olfactory perception and disturbs odor discrimination and pleasantness [60–62]. Furthermore, a recent study shows that cannabis consumption could also affect other neurotransmitter systems in olfactory structures [63]. One of the main characteristics of the CB1 receptor's activity is its bimodal activity: the cell-type of where it is expressed can lead to opposite effects (e.g., CB1 in GABAergic cells promotes satiety while in glutamatergic cells it induces hunger) [64]. This bimodal action could also be present in olfactory processes considering the pattern of expression of the CB1 receptors, but future research is needed to clarify this crucial point.

The interconnectivity between olfactory areas, together with the tight ECS-control of various types of cells and subcellular locations, makes the determination of the different roles of CB1 receptors in the olfactory system very complex and challenging. A better understanding of such interactions will result not only in a significant advance for neuroscience, but could also lead to novel human-based studies targeting specific populations. Interestingly, alterations of ECS functioning have been shown to contribute to the development of neurological and neuropsychiatric disorders in which loss of smell represents the early stages of the disease [65–68]. All of this information could provide the rationale to propose a combined use of olfactory manipulations with ECS-based pharmacotherapy to potentially treat pathological conditions.

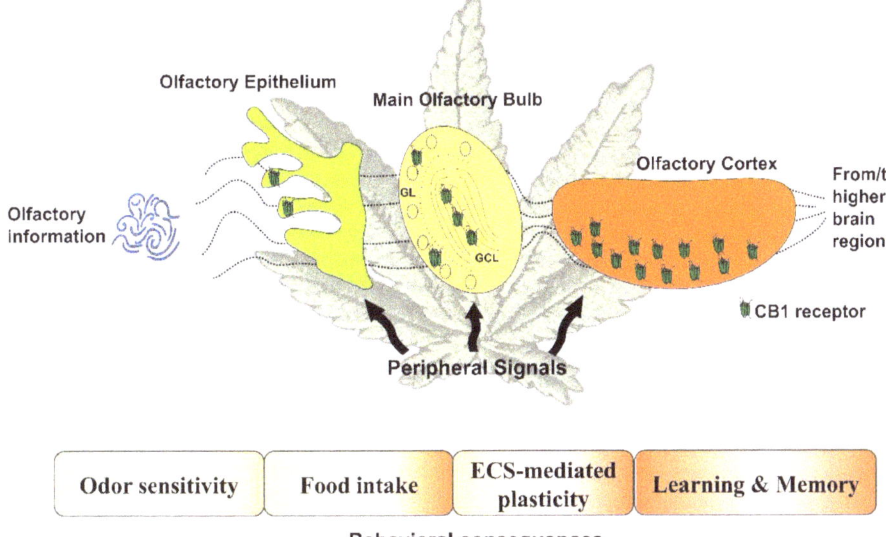

Figure 1. Main functions regulated by the cannabinoid signaling described in the primary olfactory-dependent structures. The endocannabinoid system controls various functions that depend on the olfactory areas involved. The colors in the boxes indicate which structure is involved in the associated function. The scheme also shows how the different localization of the cannabinoid type-1 (CB1) receptor potentially modulates the flow of olfactory information from early sensory coding to more complex computations, and is modulated by peripheral signals, resulting in behavioral outputs. GL: glomerular layer; GCL: granular cell layer; the olfactory cortex includes both the anterior olfactory nucleus and the piriform cortex, ECS: endocannabinoid system.

Author Contributions: Conceptualization, E.S-G.; Writing-Original Draft Preparation, G.T.; Writing-Review & Editing, P.G., G.M. and E.S.-G.; Supervision, E.S.-G. All authors have read and agreed to the published version of the manuscript.

Funding: This research was funded by Fondation pour la Recherche Médicale (FRM, FDT20170436845) (to G.T.); The Basque Government (ITI230-19), Red de Trastornos Adictivos, Instituto de Salud Carlos III (ISC-III) and European Regional Development Funds-European Union (ERDF-EU; grant RD16/0017/0012), MINECO/FEDER, UE (SAF2015-65034-R) (to P.G.); EU–FP7 (PAINCAGE, HEALTH-603191), European Research Council (Endofood, ERC–2010–StG–260515; CannaPreg, ERC-2014-PoC-640923, Micabra) (to G.M.); Ikerbasque (The Basque Foundation for Science) and MINECO (Ministerio de Economía y Competitividad) PGC2018-093990-A-I00 (to E.S.-G.).

Conflicts of Interest: The authors declare no conflict of interest.

References

1. United Nations Office on Drugs, and Crime. *World Drug Report*; United Nations Publications: New York, NY, USA, 2010.
2. Araque, A.; Castillo, P.E.; Manzoni, O.J.; Tonini, R. Synaptic functions of endocannabinoid signaling in health and disease. *Neuropharmacology* **2017**, *124*, 13–24. [CrossRef] [PubMed]
3. Kano, M.; Ohno-Shosaku, T.; Hashimotodani, Y.; Uchigashima, M.; Watanabe, M. Endocannabinoid-mediated control of synaptic transmission. *Physiol. Rev.* **2009**, *89*, 309–380. [CrossRef] [PubMed]
4. Maejima, T.; Hashimoto, K.; Yoshida, T.; Aiba, A.; Kano, M. Presynaptic inhibition caused by retrograde signal from metabotropic glutamate to cannabinoid receptors. *Neuron* **2001**, *31*, 463–475. [CrossRef]
5. Gutiérrez-Rodríguez, A.; Puente, N.; Elezgarai, I.; Ruehle, S.; Lutz, B.; Reguero, L.; Gerrikagoitia, I.; Marsicano, G.; Grandes, P. Anatomical characterization of the cannabinoid CB1 receptor in cell-type-specific mutant mouse rescue models. *J. Comp. Neurol.* **2017**, *525*, 302–318. [CrossRef]

6. Gutiérrez-Rodríguez, A.; Bonilla-Del Río, I.; Puente, N.; Gómez-Urquijo, S.M.; Fontaine, C.J.; Egaña-Huguet, J.; Elezgarai, I.; Ruehle, S.; Lutz, B.; Robin, L.M.; et al. Localization of the cannabinoid type-1 receptor in subcellular astrocyte compartments of mutant mouse hippocampus. *Glia* **2018**, *66*, 1417–1431. [CrossRef]
7. Kreitzer, A.C.; Regehr, W.G. Retrograde inhibition of presynaptic calcium influx by endogenous cannabinoids at excitatory synapses onto purkinje cells. *Neuron* **2001**, *29*, 717–727. [CrossRef]
8. Ohno-Shosaku, T.; Maejima, T.; Kano, M. Endogenous cannabinoids mediate retrograde signals from depolarized postsynaptic neurons to presynaptic terminals. *Neuron* **2001**, *29*, 729–738. [CrossRef]
9. Wilson, R.I.; Nicoll, R.A. Endogenous cannabinoids mediate retrograde signalling at hippocampal synapses. *Nature* **2001**, *410*, 588–592. [CrossRef]
10. Chevaleyre, V.; Castillo, P.E. Heterosynaptic LTD of hippocampal GABAergic synapses: A novel role of endocannabinoids in regulating excitability. *Neuron* **2003**, *38*, 461–472. [CrossRef]
11. Marsicano, G.; Wotjak, C.T.; Azad, S.C.; Bisogno, T.; Rammes, G.; Cascio, M.G.; Hermann, H.; Tang, J.; Hofmann, C.; Zieglgänsberger, W.; et al. The endogenous cannabinoid system controls extinction of aversive memories. *Nature* **2002**, *418*, 530–534. [CrossRef]
12. Bacci, A.; Huguenard, J.R.; Prince, D.A. Long-lasting self-inhibition of neocortical interneurons mediated by endocannabinoids. *Nature* **2004**, *431*, 312–316. [CrossRef]
13. Gerdeman, G.L.; Ronesi, J.; Lovinger, D.M. Postsynaptic endocannabinoid release is critical to long-term depression in the striatum. *Nat. Neurosci.* **2002**, *5*, 446–451. [CrossRef]
14. Peñasco, S.; Rico-Barrio, I.; Puente, N.; Gómez-Urquijo, S.M.; Fontaine, C.J.; Egaña-Huguet, J.; Achicallende, S.; Ramos, A.; Reguero, L.; Elezgarai, I.; et al. Endocannabinoid long-term depression revealed at medial perforant path excitatory synapses in the dentate gyrus. *Neuropharmacology* **2019**, *153*, 32–40. [CrossRef] [PubMed]
15. Howlett, A.C.; Barth, F.; Bonner, T.I.; Cabral, G.; Casellas, P.; Devane, A.; Felder, C.C.; Herkenham, M.; Mackie, K.; Martin, B.R.; et al. International union of pharmacology. XXVII. Classification of cannabinoid receptors. *Pharm. Rev.* **2002**, *54*, 161–202. [CrossRef] [PubMed]
16. Marsicano, G.; Lafenêtre, P. Roles of the endocannabinoid system in learning and memory. *Curr. Top. Behav. Neurosci.* **2009**, *1*, 201–230. [PubMed]
17. Morena, M.; Campolongo, P. The endocannabinoid system: An emotional buffer in the modulation of memory function. *Neurobiol. Learn. Mem.* **2014**, *112*, 30–43. [CrossRef] [PubMed]
18. Soria-Gómez, E.; Bellocchio, L.; Marsicano, G. New insights on food intake control by olfactory processes: The emerging role of the endocannabinoid system. *Mol. Cell. Endocrinol.* **2014**, *397*, 59–66. [CrossRef]
19. Tart, C.T. Marijuana intoxication: Common experiences. *Nature* **1970**, *226*, 701–704. [CrossRef]
20. Herkenham, M.; Lynn, A.B.; Little, M.D.; Johnson, M.R.; Melvin, L.S.; de Costa, B.R.; Rice, K.C. Cannabinoid receptor localization in brain. *Proc. Natl. Acad. Sci. USA* **1990**, *87*, 1932–1936. [CrossRef]
21. Herkenham, M.; Lynn, A.; Johnson, M.; Melvin, L.; de Costa, B.; Rice, K. Characterization and localization of cannabinoid receptors in rat brain: A quantitative in vitro autoradiographic study. *J. Neurosci.* **1991**, *11*, 563–583. [CrossRef] [PubMed]
22. Marsicano, G.; Lutz, B. Expression of the cannabinoid receptor CB1 in distinct neuronal subpopulations in the adult mouse forebrain. *Eur. J. Neurosci.* **1999**, *11*, 4213–4225. [CrossRef] [PubMed]
23. Pettit, D.A.D.; Harrison, M.P.; Olson, J.M.; Spencer, R.F.; Cabral, G.A. Immunohistochemical localization of the neural cannabinoid receptor in rat brain. *J. Neurosci. Res.* **1998**, *51*, 391–402. [CrossRef]
24. Czesnik, D.; Schild, D.; Kuduz, J.; Manzini, I. Cannabinoid action in the olfactory epithelium. *Proc. Natl. Acad. Sci. USA* **2007**, *104*, 2967–2972. [CrossRef] [PubMed]
25. Breunig, E.; Manzini, I.; Piscitelli, F.; Gutermann, B.; Di Marzo, V.; Schild, D.; Czesnik, D. The endocannabinoid 2-arachidonoyl-glycerol controls odor sensitivity in larvae of xenopus laevis. *J. Neurosci.* **2010**, *30*, 8965–8973. [CrossRef] [PubMed]
26. Breunig, E.; Czesnik, D.; Piscitelli, F.; Di Marzo, V.; Manzini, I.; Schild, D. Endocannabinoid modulation in the olfactory epithelium. In *Sensory and Metabolic Control of Energy Balance*; Meyerhof, W., Beisiegel, U., Joost, H.-G., Eds.; Springer: Berlin/Heidelberg, Germany, 2010; Volume 52, pp. 139–145.
27. Hutch, C.R.; Hillard, C.J.; Jia, C.; Hegg, C.C. An endocannabinoid system is present in the mouse olfactory epithelium but does not modulate olfaction. *Neuroscience* **2015**, *300*, 539–553. [CrossRef] [PubMed]

28. Delgado, A.; Jaffé, E.H. Acute immobilization stress modulate GABA release from rat olfactory bulb: Involvement of endocannabinoids—Cannabinoids and acute stress modulate GABA release. *Int. J. Cell Biol.* **2011**, *2011*, 1–10. [CrossRef]
29. Harvey, J.; Heinbockel, T. Neuromodulation of synaptic transmission in the main olfactory bulb. *Int. J. Environ. Res. Public Health* **2018**, *15*, 2194. [CrossRef]
30. Heinbockel, T.; Wang, Z.-J.; Brown, E.A.; Austin, P.T. Endocannabinoid signaling in neural circuits of the olfactory and limbic system. In *Cannabinoids in Health and Disease*; Meccariello, R., Chianese, R., Eds.; InTech: London, UK, 2016.
31. Pouille, F.; Schoppa, N.E. Cannabinoid receptors modulate excitation of an olfactory bulb local circuit by cortical feedback. *Front. Cell. Neurosci.* **2018**, *12*, 47. [CrossRef]
32. Soria-Gómez, E.; Bellocchio, L.; Reguero, L.; Lepousez, G.; Martin, C.; Bendahmane, M.; Ruehle, S.; Remmers, F.; Desprez, T.; Matias, I.; et al. The endocannabinoid system controls food intake via olfactory processes. *Nat. Neurosci.* **2014**, *17*, 407–415. [CrossRef] [PubMed]
33. Wang, Z.-J.; Sun, L.; Heinbockel, T. Cannabinoid receptor-mediated regulation of neuronal activity and signaling in glomeruli of the main olfactory bulb. *J. Neurosci. Off. J. Soc. Neurosci.* **2012**, *32*, 8475–8479. [CrossRef] [PubMed]
34. Wang, Z.-J.; Hu, S.S.-J.; Bradshaw, H.B.; Sun, L.; Mackie, K.; Straiker, A.; Heinbockel, T. Cannabinoid receptor-mediated modulation of inhibitory inputs to mitral cells in the main olfactory bulb. *J. Neurophysiol.* **2019**, *122*, 749–759. [CrossRef]
35. Ghosh, S.; Reuveni, I.; Zidan, S.; Lamprecht, R.; Barkai, E. Learning-induced modulation of the effect of endocannabinoids on inhibitory synaptic transmission. *J. Neurophysiol.* **2018**, *119*, 752–760. [CrossRef] [PubMed]
36. Hill, A.J.; Weston, S.E.; Jones, N.A.; Smith, I.; Bevan, S.A.; Williamson, E.M.; Stephens, G.J.; Williams, C.M.; Whalley, B.J. Δ9-Tetrahydrocannabivarin suppresses in vitro epileptiform and in vivo seizure activity in adult rats: Anticonvulsant potential of Δ9-THCV. *Epilepsia* **2010**, *51*, 1522–1532. [CrossRef] [PubMed]
37. Terral, G.; Busquets-Garcia, A.; Varilh, M.; Achicallende, S.; Cannich, A.; Bellocchio, L.; Bonilla-Del Río, I.; Massa, F.; Puente, N.; Soria-Gomez, E.; et al. CB1 Receptors in the anterior piriform cortex control odor preference memory. *Curr. Biol.* **2019**, *29*, 2455–2464. [CrossRef] [PubMed]
38. Zenko, M.; Zhu, Y.; Dremencov, E.; Ren, W.; Xu, L.; Zhang, X. Requirement for the endocannabinoid system in social interaction impairment induced by coactivation of dopamine D1 and D2 receptors in the piriform cortex. *J. Neurosci. Res.* **2011**, *89*, 1245–1258. [CrossRef] [PubMed]
39. Bhutani, S.; Howard, J.D.; Reynolds, R.; Zee, P.C.; Gottfried, J.; Kahnt, T. Olfactory connectivity mediates sleep-dependent food choices in humans. *eLife* **2019**, *8*, e49053. [CrossRef]
40. Terral, G.; Varilh, M.; Cannich, A.; Massa, F.; Ferreira, G.; Marsicano, G. Synaptic Functions of Type-1 Cannabinoid receptors in inhibitory circuits of the anterior piriform cortex. *Neuroscience* **2020**, *433*, 121–131. [CrossRef]
41. Busquets-Garcia, A.; Oliveira da Cruz, J.F.; Terral, G.; Zottola, A.C.P.; Soria-Gómez, E.; Contini, A.; Martin, H.; Redon, B.; Varilh, M.; Ioannidou, C.; et al. Hippocampal CB1 receptors control incidental associations. *Neuron* **2018**, *99*, 1247–1259. [CrossRef]
42. Soria-Gómez, E.; Busquets-Garcia, A.; Hu, F.; Mehidi, A.; Cannich, A.; Roux, L.; Louit, I.; Alonso, L.; Wiesner, T.; Georges, F.; et al. Habenular CB1 receptors control the expression of aversive memories. *Neuron* **2015**, *88*, 306–313. [CrossRef]
43. Laviolette, S.R.; Grace, A.A. Cannabinoids potentiate emotional learning plasticity in neurons of the medial prefrontal cortex through basolateral amygdala inputs. *J. Neurosci.* **2006**, *26*, 6458–6468. [CrossRef]
44. Back, F.P.; Carobrez, A.P. Periaqueductal gray glutamatergic, cannabinoid and vanilloid receptor interplay in defensive behavior and aversive memory formation. *Neuropharmacology* **2018**, *135*, 399–411. [CrossRef]
45. O'Doherty, J.; Rolls, E.T.; Francis, S.; Bowtell, R.; McGlone, F.; Kobal, G.; Renner, B.; Ahne, G. Sensory-specific satiety-related olfactory activation of the human orbitofrontal cortex. *Neuroreport* **2000**, *4*, 893–897. [CrossRef] [PubMed]
46. Pager, J.; Giachetti, I.; Holley, A.; Le Magnen, J. A selective control of olfactory bulb electrical activity in relation to food deprivation and satiety in rats. *Physiol. Behav.* **1972**, *9*, 573–579. [CrossRef]
47. Di Marzo, V.; Matias, I. Endocannabinoid control of food intake and energy balance. *Nat. Neurosci.* **2005**, *8*, 585–589. [CrossRef] [PubMed]

48. Palouzier-Paulignan, B.; Lacroix, M.-C.; Aime, P.; Baly, C.; Caillol, M.; Congar, P.; Julliard, A.K.; Tucker, K.; Fadool, D.A. Olfaction under metabolic influences. *Chem. Senses* **2012**, *37*, 769–797. [CrossRef] [PubMed]
49. Zimmer, A. *Genetic Manipulation of the Endocannabinoid System*; Springer: New York, NY, USA, 2015.
50. Pinching, A.J.; Powell, T.P.S. The neuropil of the glomeruli of the olfactory bulb. *J. Cell Sci.* **1971**, *9*, 347–377.
51. Piredda, S.; Gale, K. A crucial epileptogenic site in the deep prepiriform cortex. *Nature* **1985**, *317*, 623–625. [CrossRef]
52. Marsicano, G.; Goodenough, S.; Monory, K.; Hermann, H.; Eder, M.; Cannich, A.; Azad, S.C.; Cascio, M.G.; Gutiérrez, S.O.; van der Stelt, M.; et al. CB1 cannabinoid receptors and on-demand defense against excitotoxicity. *Science* **2003**, *302*, 84–88. [CrossRef]
53. Monory, K.; Massa, F.; Egertová, M.; Eder, M.; Blaudzun, H.; Westenbroek, R.; Kelsch, W.; Jacob, W.; Marsch, R.; Ekker, M.; et al. The endocannabinoid system controls key epileptogenic circuits in the hippocampus. *Neuron* **2006**, *51*, 455–466. [CrossRef]
54. Neville, K.R.; Haberly, L.B. Olfactory cortex. In *The Synaptic Organization of the Brain*, 5th ed.; Shepherd, G.M., Ed.; Oxford University Press: Oxford, UK, 2004; Volume 8, pp. 415–454.
55. Sigel, E.; Baur, R.; Racz, I.; Marazzi, J.; Smart, T.G.; Zimmer, A.; Gertsch, J. The major central endocannabinoid directly acts at GABAA receptors. *Proc. Natl. Acad. Sci. USA* **2011**, *108*, 18150–18155. [CrossRef]
56. Gottfried, J.A.; O'Doherty, J.; Dolan, R.J. Appetitive and aversive olfactory learning in humans studied using event-related functional magnetic resonance imaging. *J. Neurosci.* **2002**, *22*, 10829–10837. [CrossRef]
57. Buchanan, T.W. A specific role for the human amygdala in olfactory memory. *Learn. Mem.* **2003**, *10*, 319–325. [CrossRef] [PubMed]
58. Schoenbaum, G.; Eichenbaum, H. Information coding in the rodent prefrontal cortex. I. Single-neuron activity in orbitofrontal cortex compared with that in pyriform cortex. *J. Neurophysiol.* **1995**, *74*, 733–750. [CrossRef] [PubMed]
59. Nogi, Y.; Ahasan, M.M.; Murata, Y.; Taniguchi, M.; Sha, M.F.R.; Ijichi, C.; Yamaguchi, M. Expression of feeding-related neuromodulatory signalling molecules in the mouse central olfactory system. *Sci. Rep.* **2020**, *10*, 890. [CrossRef]
60. Lotsch, J.; Hummel, T. Cannabinoid-related olfactory neuroscience in mice and humans. *Chem. Senses* **2015**, *40*, 3–5. [CrossRef] [PubMed]
61. Walter, C.; Oertel, B.G.; Ludyga, D.; Ultsch, A.; Hummel, T.; Lötsch, J. Effects of 20 mg oral Δ^9-tetrahydrocannabinol on the olfactory function of healthy volunteers: Effects of Δ^9-tetrahydrocannabinol on olfaction. *Br. J. Clin. Pharm.* **2014**, *78*, 961–969. [CrossRef] [PubMed]
62. Walter, C.; Oertel, B.G.; Felden, L.; Nöth, U.; Vermehren, J.; Deichmann, R.; Lötsch, J. Effects of oral Δ^9-tetrahydrocannabinol on the cerebral processing of olfactory input in healthy non-addicted subjects. *Eur. J. Clin. Pharm.* **2017**, *73*, 1579–1587. [CrossRef]
63. Galindo, L.; Moreno, E.; López-Armenta, F.; Guinart, D.; Cuenca-Royo, A.; Izquierdo-Serra, M.; Xicota, L.; Fernandez, C.; Menoyo, E.; Fernández-Fernández, J.M.; et al. Cannabis users show enhanced expression of CB1-5HT2A receptor heteromers in olfactory neuroepithelium cells. *Mol. Neurobiol.* **2018**, *55*, 6347–6361. [CrossRef]
64. Bellocchio, L.; Lafenêtre, P.; Cannich, A.; Cota, D.; Puente, N.; Grandes, P.; Chaouloff, F.; Piazza, P.V.; Marsicano, G. Bimodal control of stimulated food intake by the endocannabinoid system. *Nat. Neurosci.* **2010**, *13*, 281–283. [CrossRef]
65. Basavarajappa, B.S.; Shivakumar, M.; Joshi, V.; Subbanna, S. Endocannabinoid system in neurodegenerative disorders. *J. Neurochem.* **2017**, *142*, 624–648. [CrossRef]
66. Godoy, M.; Voegels, R.; Pinna, F.; Imamura, R.; Farfel, J. Olfaction in neurologic and neurodegenerative diseases: A literature review. *Int. Arch. Otorhinolaryngol.* **2014**, *19*, 176–179. [PubMed]
67. Philpott, C.M.; Boak, D. The impact of olfactory disorders in the United Kingdom. *Chem. Senses* **2014**, *39*, 711–718. [CrossRef] [PubMed]
68. Yin, A.; Wang, F.; Zhang, X. Integrating endocannabinoid signaling in the regulation of anxiety and depression. *Acta Pharm. Sin.* **2018**, *40*, 336–341. [CrossRef] [PubMed]

Article

Excitability of Neural Activity is Enhanced, but Neural Discrimination of Odors is Slightly Decreased, in the Olfactory Bulb of Fasted Mice

Jing Wu [1,†], Penglai Liu [1,†], Fengjiao Chen [1], Lingying Ge [2], Yifan Lu [2] and Anan Li [1,*]

1 Jiangsu Key Laboratory of Brain Disease and Bioinformation, Research Center for Biochemistry and Molecular Biology, Xuzhou Medical University, Xuzhou 221004, China; jingwu@stu.xzhmu.edu.cn (J.W.); penglailiu@gmail.com (P.L.); fengjiaochen@xzhmu.edu.cn (F.C.)

2 The Second Clinical Medical College, Xuzhou Medical University, Xuzhou 221004, China; gelingying@outlook.com (L.G.); lyf00614@outlook.com (Y.L.)

* Correspondence: anan.li@xzhmu.edu.cn; Tel.: +86-516-83262621

† These authors contributed equally to the work.

Received: 30 March 2020; Accepted: 14 April 2020; Published: 16 April 2020

Abstract: Olfaction and satiety status influence each other: cues from the olfactory system modulate eating behavior, and satiety affects olfactory abilities. However, the neural mechanisms governing the interactions between olfaction and satiety are unknown. Here, we investigate how an animal's nutritional state modulates neural activity and odor representation in the mitral/tufted cells of the olfactory bulb, a key olfactory center that plays important roles in odor processing and representation. At the single-cell level, we found that the spontaneous firing rate of mitral/tufted cells and the number of cells showing an excitatory response both increased when mice were in a fasted state. However, the neural discrimination of odors slightly decreased. Although ongoing baseline and odor-evoked beta oscillations in the local field potential in the olfactory bulb were unchanged with fasting, the amplitude of odor-evoked gamma oscillations significantly decreased in a fasted state. These neural changes in the olfactory bulb were independent of the sniffing pattern, since both sniffing frequency and mean inhalation duration did not change with fasting. These results provide new information toward understanding the neural circuit mechanisms by which olfaction is modulated by nutritional status.

Keywords: olfactory bulb; nutritional status; in vivo electrophysiological recording; odor representation

1. Introduction

Food intake is a complex process in which both homoeostatic regulation and hedonic sensations are critically involved. Most sensory systems influence food detection and consumption [1–4]. However, of all the sensory modalities, olfaction contributes the most to the hedonic evaluation of a food and its eventual possible consumption [1,2,5]. Conversely, metabolic states such as fasting or satiation have been reported to increase or decrease olfactory detection and discrimination in both humans and rodents [6–8]. Although it is well known that olfaction and satiety status influence each other, the underlying neural mechanisms are largely unknown.

The representation of odor information is rather complex regarding the need to process parallel input from different olfactory receptors and trace amine-associated receptors expressed from more than 1000 genes in rodents [9–11]. The olfactory bulb (OB) is the first relay station and processing hub in the olfactory system. Recent studies have demonstrated that the OB plays a key role in the representation of odor identity, intensity, and timing [12–15]. In the OB, mitral/tufted cells (M/Ts) are the main output neurons that send the processed neural signals to higher olfactory centers for further information processing. Thus, factors that influence the activity of M/Ts or their ability to represent odors may cause deficiencies in olfactory function [12]. Therefore, it is important to decipher how nutritional status affects neural activity and odor representation in M/Ts.

Pioneering work performed by Pager found that more multiple units recorded from rat OB showed excitatory responses to a food odor when the animal was in a fasted state compared with a satiated state [16]. This finding was further supported by another study in which spikes were recorded from single units [17]. However, these studies only compared the number of units showing different response types; further in-depth analysis of how nutritional status influences neural discrimination of odors in M/Ts is lacking. Furthermore, whether the change in neural response in different nutritional states is dependent on changes in sniffing also remains unknown.

In the current study, we first tested how nutritional status modulates single-unit activity and neural discrimination of odors in M/Ts recorded from awake, head-fixed mice. We then investigated ongoing and odor-evoked local field potential (LFP) responses under different fasting states. Finally, we asked whether changes in the neural activity in the OB are related to changes in the animal's sniffing pattern. We found that the excitability of neural activity is enhanced, but the neural discrimination of odors is slightly decreased in the OB of fasted mice.

2. Materials and Methods

2.1. Animals

Male eight-week-old C57BL/6J mice were used as experimental subjects and were housed under a 12 h light/dark cycle with food and water ad libitum. Normally, four or five mice were placed in one cage, but mice were housed individually after surgery for at least one week for recovery before further experiments. All experimental procedures complied with the animal care standards of the Xuzhou Medical University Institutional Animal Care and Use Committee (SYXK2015-0030).

2.2. Odorants and Preference/Avoidance Behavioral Test

Odorants were applied in three groups: neutral odorants (isoamyl acetate, 2-heptanone), appetitive odorants (peanut butter, food pellets), and aversive odorants (2,4,5-trimethylthiazole, peppermint oil). The odorants were dissolved in mineral oil at 40% v/v dilution; peanut butter was used in their original states and food was dissolved in saline until the saline was saturated. For each animal, all six odors were tested. The interval between two odors was at least two days. The odors were presented in the order of appetitive odorants (Food, Peanut butter), neutral odorants (Isoamyl acetate, 2-heptanone), and aversive odorants (2,4,5-trimethylthiazole, peppermint oil). During testing, the odorant (50 μL), peanut butter (1 g), food (1 g), or mineral oil (50 μL) was placed on a filter paper in a dish (60 mm × 15 mm) and covered with cage bedding. A custom-designed test chamber (45 cm × 35 cm × 25 cm) with two equally sized compartments was used. Before the test, the mouse was placed into the chamber with empty dishes for 10 min to habituate to the environment. Then, preference for/avoidance of odorants was tested by exposing mice to the two compartments for 10 min, with an odorant in one compartment and mineral oil in the other compartment (the locations of the odorant and mineral oil were randomized) (Figure 1A). Odor preference or avoidance was reflected by the animal's movement trace—the time spent in each chamber was calculated automatically by a computerized recording system.

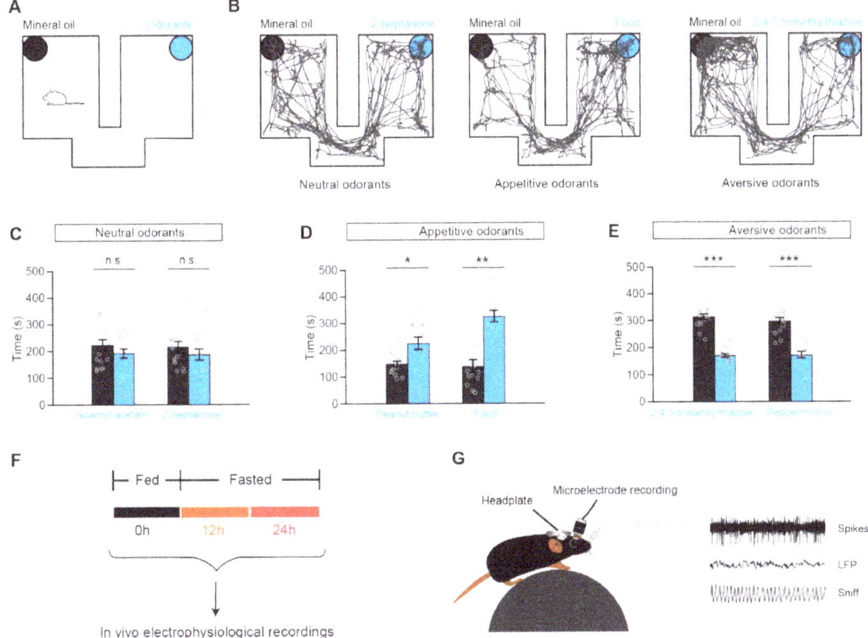

Figure 1. Paradigm for the preference/avoidance test and in vivo recordings. (**A**) Schematic of the preference/avoidance test. (**B**) Representative movement traces illustrating three different behaviors in C57BL/6J mice (left, a neutral odorant: 2-heptanone; middle, an appetitive odorant: food; right, an aversive odorant: 2,4,5-trimethylthiazole). (**C–E**) Quantification of time spent by mice in the chamber with the odorant (blue) versus mineral oil (black), for each type of odorant. n = 15 for each pair. (**C**) Mice showed neither preference nor avoidance for neutral odorants compared with mineral oil (paired t-tests: isoamyl acetate vs. mineral oil, $t_{(14)} = -0.83$, $p = 0.42$; 2-heptanone vs. mineral oil, $t_{(14)} = 0.79$, $p = 0.44$). (**D**) Mice showed preference for appetitive odorants compared with mineral oil (paired t-tests: peanut butter vs. mineral oil, $t_{(14)} = 2.82$, $p = 0.013$; Wilcoxon signed-rank test: Food vs. mineral oil, $z = -2.78$, $p = 0.0053$). (**E**) Mice showed avoidance for aversive odorants compared with mineral oil (paired t-tests: 2,4,5-trimethylthiazole vs. mineral oil, $t_{(14)} = -8.58$, $p < 0.001$; peppermint oil vs. mineral oil, $t_{(14)} = -5.80$, $p < 0.001$). (**F**) Diagram of the in vivo electrophysiological recordings. The recordings repeated at different metabolism states, including fasted for 0 h, 12 h, and 24 h. (**G**) Schematic representation of the methods for in vivo electrophysiological recordings in awake, head-fixed mice. n.s., not significant. * $p < 0.05$, ** $p < 0.01$, *** $p < 0.001$.

2.3. Microelectrode Implantation

The microelectrodes (16-channel, Jiangsu Brain Medical Technology Co. Ltd, Nanjing, China) were implanted into a specific region of the brain as previously described [18,19]. Briefly, mice were anesthetized with pentobarbital sodium (90 mg/kg body weight, i.p.) and positioned in a stereotaxic frame. Eye ointment was applied to the eyes. The skull surface (from the midline of the orbits to the midpoint between the ears) was exposed, and a hole was drilled above the right OB for microelectrode implantation (anterior-posterior (AP): +4.0 mm; medial-lateral (ML): +1.0 mm). Then, the microelectrodes were positioned and lowered through the drilled holes until they reached the OB mitral cell layer at an average depth between 1.8 mm and 2.5 mm. A custom-designed head plate was attached to the skull with small screws and dental acrylic to enable head fixation during recordings. The body temperature of mice was maintained at 37 ± 0.5 °C throughout the surgery.

2.4. Spike and LFP Recordings

The recordings were initiated after the mice had recovered from surgery, as in previous studies [18–21]. Briefly, awake mice were head-fixed with two horizontal bars and were able to maneuver on an air-supported free-floating Styrofoam ball (Thinkerbiotech, Nanjing, China). For spike recordings, the signals from the microelectrodes were sent to a headstage, amplified by a 16-channel amplifier (Plexon DigiAmp (Plexon Inc, Dallas, TX, USA); bandpass filtered at 300–5000 Hz, 2000× gain), and sampled at 40 kHz by a Plexon Omniplex recording system. For LFP recordings, LFP signals were amplified (2000× gain, Plexon DigiAmp), filtered at 0.1–300 Hz, and sampled at 1 kHz. Spikes or LFP signals together with odor stimulation event markers were recorded via the same Plexon Omniplex recording system. The fasting started at 21:00 and finished at 21:00 the next day. Recordings were repeated on the same mice under different nutritional states: satiety, fasted for 12 h, and fasted for 24 h.

2.5. Odorant Presentation during Electrophysiological Recordings

The three sets of odorants described above were also used during electrophysiological recordings. Peanut butter was mixed with mineral oil at a 10% m/v dilution and food was dissolved in saline as described above. The other odorants were dissolved in mineral oil at 1% v/v dilution. During the odor delivery period, an odor delivery system (Thinkerbiotech, Nanjing, China) was used, as previously described [20,21]. There were 15 trials for each odorant. The six odorants were presented in a pseudo-randomized order, with no more than two successive presentations of the same odor. Each odor was delivered to the animal for 2 s with an inter-stimulus interval of 20 s. All six odorants were presented passively; the mice were not required to respond to the odors presented.

2.6. Measurement of Sniffing Parameters

The sniffing patterns of mice during electrophysiological recordings were recorded continuously by placing a cannula into one nasal cavity and connecting it to an airflow pressure sensor (Model No. 24PCEFA6G(EA), 0–0.5 psi, Honeywell) Surgical implantation of the nasal cannula was performed as described in previous studies [22,23]. The pressure transient signals were amplified (100× gain, Plexon DigiAmp) and sampled at 1 kHz by a Plexon Omniplex recording system. A sniff was defined as the point of transition from exhalation to inhalation.

2.7. Data Analysis

2.7.1. Olfactory Preference/Avoidance Test

Preference time and avoidance time during the 10-min test were measured from the recorded videos using MATLAB. The test cage was divided into two compartments of equal area. Preference time was defined as the time spent in the compartment with a filter paper scented with peanut butter or food, and avoidance time was defined as the time spent in a compartment without a filter paper scented with 2,4,5-trimethylthiazole or peppermint oil [24].

Offline spike sorting and statistics of single-cell spiking data: When we did spike sorting, all the files were put together to make sure signals recorded at different stages were from same units. Similar to previous studies, single units were sorted and identified with principal components analysis in Offline Sorter V4 software [18,19]. In addition, we performed further analysis on all recorded unit to double check that the spikes were from the same units (repeated measures ANOVA on the spike amplitude and half-width). To generate the peristimulus time histogram (PSTH), spikes 2 s before and 4 s after the onset of odor stimulation were extracted for each trial and the spike firing rate was averaged over 100-ms bins (Figure 2E). The spontaneous firing rate (during the 2 s before odor stimulation) and the odor-evoked firing rate (during the 2 s after odor stimulation) were calculated by averaging the frequency of spikes during these 2 s periods. To test whether an odor evoked a significant response, we used a paired t-test to compare the baseline firing rate with the odor-evoked firing rate across all the trials for each cell–odor pair. If the p value was >0.05, the cell–odor pair was defined as nonresponsive.

If the p value was <0.05, the cell–odor pair was defined as responsive and was further categorized as excitatory (if the odor-evoked firing rate was higher than the baseline firing rate) or inhibitory (if the odor-evoked firing rate was lower than the baseline firing rate).

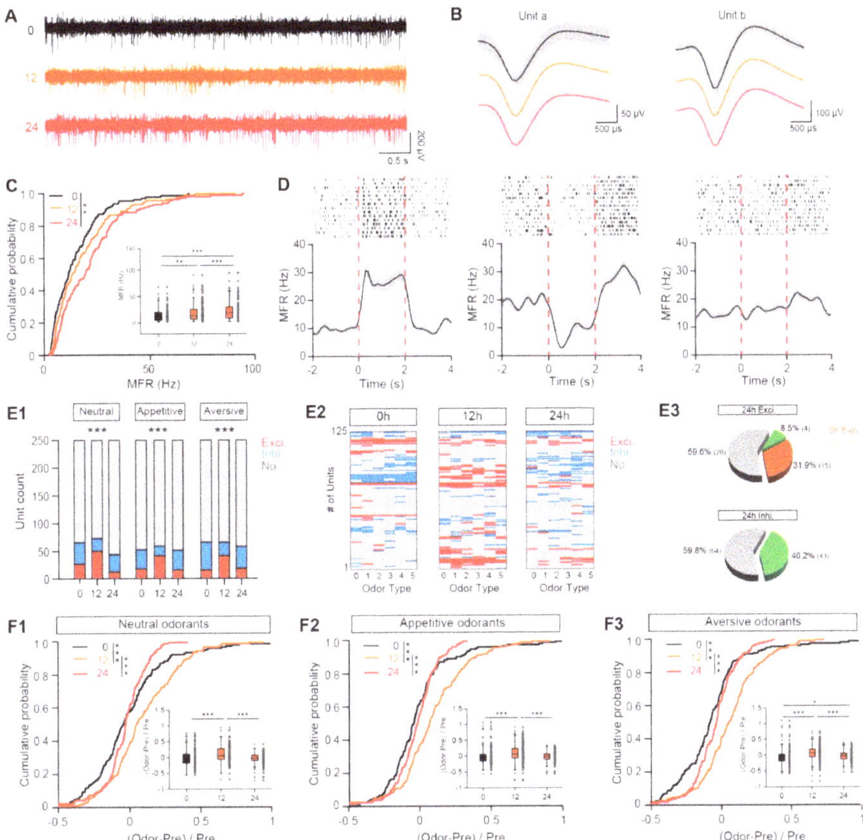

Figure 2. The firing of M/Ts recorded from the OB changes under different nutritional states. (**A**,**B**) Representative raw spike traces (**A**) and spike waveforms (**B**) recorded from a mouse fasted for 0 h (black), 12 h (orange), and 24 h (red) respectively. (**C**) Cumulative probability and box-and-whisker plot showing the spontaneous firing rate under different fasting states. Each gray circle represents the mean firing rate of a single unit. (Cumulative probability: Two-sample *K-S* test, 0 h vs. 12 h, $p = 0.31$, 0 h vs. 24 h, $p = 0.0012$, 12 h vs. 24 h, $p = 0.073$. Box-and-whisker plot: Friedman's test, $\chi^2_{(2,248)} = 65.10$, $p < 0.001$, 0 h vs. 12 h, $p = 0.0023$, 0 h vs. 24 h, $p < 0.001$, 12 h vs. 24 h, $p < 0.001$). (**D**) Three examples of firing induced by isoamyl acetate when mice were in a satiated state. From left to right: an excitatory response, an inhibitory response, and no response. Top, raster plot. Bottom, peristimulus histograms for the firing rate, smoothed with a Gaussian filter with a standard deviation of 1500 ms. The red dotted lines indicate the period of odor stimulation. Error bars show the standard error of the mean (SEM). (**E**) The changes in the neural responses to odors under different fasting conditions. (**E1**). Stacked bar plots Neutral odorants (0 h, Excitory: 27/250, Inhibitory: 39/250, No response: 184/250; 12 h, Excitory:

50/250, Inhibitory: 23/250, No response: 177/250; 24 h, Excitatory: 12/250, Inhibitory: 32/250, No response: 206/250). Chi-Square Tests: $\chi^2_{(4)} = 31.23$, $p = 0.000003$, Exci. 0 h vs. Exci. 12 h, $p < 0.05$, Exci. 0 h vs. Exci. 24 h, $p < 0.05$, Exci. 12 h vs. Exci. 24 h, $p < 0.05$. Appetitive odorants (0 h, Excitory: 18/250, Inhibitory: 35/250, No response: 197/250; 12 h, Excitory: 42/250, Inhibitory: 17/250, No response: 191/250; 24 h, Excitory: 16/250, Inhibitory: 36/250, No response: 198/250). Chi-Square Tests: $\chi^2_{(4)} = 24.47$, $p = 0.000064$, Exci. 0 h vs. Exci. 12 h, $p < 0.05$, Exci. 12 h vs. Exci. 24 h, $p < 0.05$. Aversive odorants (0 h, Excitory: 16/250, Inhibitory: 50/250, No response: 184/250; 12 h, Excitory: 42/250, Inhibitory: 24/250, No response: 184/250; 24 h, Excitory: 19/250, Inhibitory: 39/250, No response: 192/250). Chi-Square Tests: $\chi^2_{(4)} = 25.04$, $p = 0.000049$, Exci. 0 h vs. Exci. 12 h, $p < 0.05$, Exci. 12 h vs. Exci. 24 h, $p < 0.05$. (**E2**). Pseudocolor plots represent odor-evoked responses for each unit at different metabolism states, including fasted for 0 h, 12 h, and 24 h. (**E3**). Among the number of unit-odor pairs with excitatory response (Top) or inhibitory response (Bottom) at fasted for 24 h, the percentages for unit-odor pairs showing excitatory (orange), inhibitory (green) or no response (gray) at fasted for 0 h. (**F**) Cumulative probability and box-and-whisker plots of the odor-evoked firing rate. (**F1**). Cumulative probability: Two-sample *K-S* test: 0 h vs. 12 h, $p < 0.001$, 0 h vs. 24 h, $p = 0.082$, 12 h vs. 24 h, $p < 0.001$. Box-and-whisker plot: Friedman's test: $\chi^2_{(2,310)} = 35.81$, $p < 0.001$, 0 h vs. 12 h, $p < 0.001$, 12 h vs. 24 h, $p < 0.001$. (**F2**). Cumulative probability: Two-sample *K-S* test: 0 h vs. 12 h, $p < 0.001$, 0 h vs. 24 h, $p = 0.24$, 12 h vs. 24 h, $p < 0.001$. Box-and-whisker plot: Friedman's test: $\chi^2_{(2,310)} = 17.78$, $p = 0.00014$, 0 h vs. 12 h, $p = 0.00085$, 12 h vs. 24 h, $p = 0.00068$. (**F3**). Cumulative probability: Two-sample *K-S* test: 0 h vs. 12 h, $p < 0.001$, 0 h vs. 24 h, $p = 0.023$, 12 h vs. 24 h, $p < 0.001$. Box-and-whisker plot: Friedman's test: $\chi^2_{(2,310)} = 30.36$, $p < 0.001$, 0 h vs. 12 h, $p < 0.001$, 12 h vs. 24 h, $p < 0.001$. * $p < 0.05$, ** $p < 0.01$, *** $p < 0.001$.

2.7.2. Analysis of LFP Signals

Programs written in MATLAB were used to analyze the LFP signals. Raw data 2 s prior to the onset of odor stimulation were used to represent the ongoing baseline LFP activity. A time–frequency transformation was performed on this 2-s window. For odor-evoked responses, the data 2 s prior to and 4 s after the onset of odor stimulation were selected for presentation and further analysis. Similar to previous studies [20,21], we divided the LFP signals into four frequency bands: theta (2–12 Hz), beta (15–35 Hz), low gamma (36–65 Hz), and high gamma (66–95 Hz). However, we focused only on the beta and high gamma bands in our analysis since odors usually evoke strong and reliable responses within these two frequency bands. Spectral power was computed using MATLAB's STFT method (The MathWorks). For each trial, the baseline was normalized to 1, and all the trials for each odor were averaged for further analysis.

2.7.3. Receiver Operating Characteristic (ROC) Analysis

Receiver operating characteristics (ROCs) were used to assess the classification of responses evoked by odor pairs, and were estimated using the roc function in MATLAB. Mean firing rate during odor stimuli with a 2-s bin was utilized in ROC analysis. The area under the ROC (auROC) is a nonparametric measure of the discriminability of two distributions. We used the auROC to assess the classification of the two odors within an odor pair. An auROC curve is defined from 0.5 to 1.0. A value of 0.5 indicates completely overlapping distributions, whereas a value of 1 predicts perfect discriminability [19].

2.7.4. Statistics

Data were analyzed in MATLAB. The Gaussian distribution of the data was assessed using the Anderson–Darling test. If the data sets were normally distributed, the data were tested for significance using a paired *t*-test (two related samples) or one-way repeated measures ANOVA (>2 related samples). If the data sets were non-normally distributed, the data were tested for significance using the Wilcoxon signed-rank test (two related samples), Friedman's test (>2 related samples), or the Kruskal-Wallis test

(>2 unrelated samples). Tukey post-hoc tests were used to directly assess group differences following ANOVA where appropriate. Where boxplots are used to represent the data, the median is plotted as a line within a box formed by the 25th (q_1) and 75th (q_3) percentiles. Points are drawn as outliers if they are larger than $q_3 + w \times (q_3 - q_1)$ or smaller than $q_1 - w \times (q_3 - q_1)$.

3. Results

To test the odor responses of OB neurons under satiated and fasted states, different types of odorants were used. Isoamyl acetate and 2-heptanone were used as neutral odorants, peanut butter [24–26] and food odor were used as appetitive odorants, and 2,4,5-trimethylthiazole and peppermint oil [27] were used as aversive odorants. To confirm that the mice had a preference for the appetitive odorants and avoided the aversive odorants, we performed a preference/avoidance test (Figure 1A). An example is illustrated in Figure 1B: although the mouse had no preference or avoidance for 2-heptanone versus mineral oil, it demonstrated a preference for food odor and avoidance of 2,4,5-trimethylthiazole. Further analysis across all the mice tested demonstrated that isoamyl acetate/2-heptanone, peanut butter/food odor, and 2,4,5-trimethylthiazole/peppermint oil were neutral, appetitive, and aversive odorants for mice, respectively (Figure 1C–E).

To compare the neural activity and sniffing patterns in satiated and fasted states, signals were recorded before the removal of food, and 12 hours and 24 hours after the removal of food (Figure 1F). Sniffing signals and neural activity, including spikes and LFP, were recorded simultaneously in awake, head-fixed mice (Figure 1G).

3.1. *Baseline Firing Rate and Odor-Evoked Responses are Both Enhanced in a Fasted State*

First, we investigated the spontaneous neural activity of single M/Ts under different nutritional states. Extracellular microelectrodes were placed into the mitral cell layer and single M/T units were isolated and sorted as described previously [18,19,23]. As in previous studies, we observed strong spontaneous firing of M/Ts in awake mice (Figure 2A). Figure 2B shows examples of two single M/Ts sorted from microelectrode recordings. The shapes of these units were similar across satiated and fasted states, indicating that the signals were likely collected from the same units under different states. We performed further analysis on all recorded unit to double check that the spikes were from the same units (repeated measures ANOVA on the spike amplitude and half-width). The data showed that signals recorded at different stages were not significantly different (Friedman's test, for amplitude, $x^2_{(2,248)} = 4.501$, $p = 0.11$; for half-width, $x^2_{(2,248)} = 4.36$, $p = 0.11$), indicating they were likely from same units. Compared with the satiated state, the spontaneous firing of M/Ts was significantly increased at 12 hours after removal of food, and further increased at 24 hours after removal of food (Figure 2C). In addition, we also performed control experiment in which we recorded data at different time points (0 h, 12 h, 24 h), but the food was not removed. The data showed that spontaneous firing of M/Ts at different time point with food were not significantly different (Two-sample *K-S* test, 0 h vs. 12 h, $p = 0.18$, 0 h vs. 24 h, $p = 0.10$, 12 h vs. 24 h, $p = 0.99$; Friedman's test, $x^2_{(2,248)} = 0.4$, $p = 0.82$).

Next, we investigated how fasting affects the odor-evoked responses of M/Ts. Consistent with the findings from previous studies, M/Ts showed both excitatory and inhibitory responses to odor stimulation in the satiated state (Figure 2D). Compared with the satiated state, the number of units showing excitatory responses was significantly increased 12 h after the removal of food and the number of units showing inhibitory responses was significantly decreased, for all three types of odorant (neutral, appetitive, and aversive) (Figure 2(E1)). Interestingly, this tendency was not observed 24 h after the removal of food (Figure 2(E1)). Control experiment showed that excitatory responses at different time point with food was not significantly different, for all three types of odorant (Chi-Square Tests, all $p > 0.05$). Since the number of responsive units under satiated state was similar with 24 h after the removal of food, this raises the question that whether they were the same set of units. We provided further presentations of the odor-evoked responses (Figure 2(E2,E3)). Interestingly, we found that most of the units showing excitatory responses under over-fasted state (24 h after removal of food) were not the same units showing excitatory responses under satiated state, indicating that the neural connectivity was reconfigured under over-fasted state (Figure 2(E2,E3)).

To further compare the amplitude of odor-evoked responses under different states, we analyzed the normalized odor response. We found that, compared with the satiated state, the odor response was increased 12 h after the removal of food for all odorants tested, and recovered 24 h after the removal of food (Figure 2F). This finding is consistent with the changes in the number of responsive units under different fasting states. Together, these results from single unit recordings indicate that the excitability of M/Ts is enhanced in fasted mice.

3.2. Neural Discrimination of Odors is Slightly Decreased in the OB of Fasted Mice

The significant difference in the excitability of M/Ts in satiated and fasted states raises the question of whether odor discrimination by single M/Ts is different under these two states. To investigate this, we characterized the ability of single-unit M/Ts to discriminate the odors. To compare the classification of odor-evoked responses under different nutritional states, we calculated the receiver operating characteristics (ROC) [19,28]. Figure 3A shows two example ROC plots—whereas the ROC curves were similar under different states for the odor pair of peanut butter and food odor (Figure 3A, left), the ROC curves were different under different states for the odor pair of food odor and 2,4,5-trimethylthiazole (Figure 3A, right). ROC analysis of all animals showed that the auROC values for the peanut butter/food odor pair were similar for satiated and fasted states (Figure 3B, left) but the auROC values for the food odor/2,4,5-trimethylthiazole pair were significantly different in different states (Figure 3B, right). We analyzed all odor pairs and found that most of the odor pairs showed smaller auROC values under fasting (Figure 3C,D, left), although no specific odor pair other than food odor/2,4,5-trimethylthiazole had a significant difference in auROC values between the satiated and fasted states (Figure 3C,D, right). The cumulative probability of auROCs analysis of all odor pairs from all animals showed that the auROC values under the satiated state were significantly larger than under the fasted states (Figure 3E) both 12 h and 24 h after the removal of food. Thus, these results indicate that, compared with the satiated state, there was a tendency that the neural discrimination was slightly decreased in the OB of fasted mice under awake, head-fixed conditions.

3.3. Odor-Evoked Gamma Responses are Decreased in a Fasted State

Whereas single-unit spiking reflects the activity in a single M/T cell, oscillations in the LFP reflect the neural activity in the population of cells surrounding the recording site [29]. Oscillations recorded from the OB contain important information relating to the chemical properties of odors, olfactory learning, and odor discrimination [30,31]. Thus, we next compared the LFP signals in awake, head-fixed mice in satiated and fasted states. As in previous studies, the raw LFP signals were divided into different frequency bands: theta, 2–12 Hz; beta, 15–35 Hz; low gamma, 36–65 Hz; and high gamma, 66–95 Hz (Figure 4A). No significant differences were found between the satiated and fasted states for any frequency band of the ongoing, baseline LFP (Figure 4B).

Next, we investigated whether odor-evoked LFP responses differed with fasting state in awake, head-fixed mice. Figure 5A–C show the LFP response to isoamyl acetate in a single mouse across different fasting states. In all states, there was a strong beta response to the odor and a high gamma response. However, although the amplitude of the isoamyl acetate-induced beta-band response was similar in the fasted state and the satiated state, the amplitude of the isoamyl acetate-induced high-gamma response decreased with fasting (Figure 5B,C). This phenomenon was also observed for other odorants (e.g., Figure 5D–I).

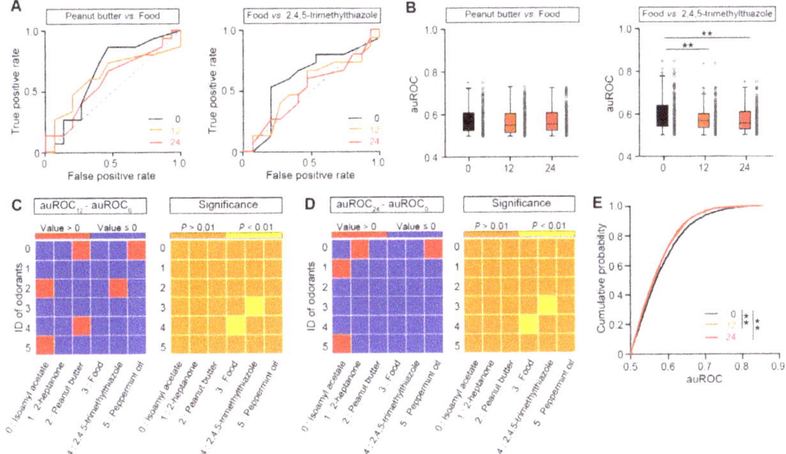

Figure 3. Nutritional status influence odor representation in M/Ts. (**A**) Example receiver operating characteristic (ROC) plots of the neural responses to peanut butter vs. food (left) and food vs. 2,4,5-trimethylthiazol (right) when mice had been fasted for 0 h (black), 12 h (orange), or 24 h (red). (**B**) Comparison of areas under the ROC (auROCs) for the neural responses to peanut butter vs. food (left) and food vs. 2,4,5-trimethylthiazol (right). Each gray circle represents the auROC value for a single unit. Left. Friedman's test: $\chi^2_{(2,248)} = 0.36$, $p = 0.83$. Right. Friedman's test: $\chi^2_{(2,248)} = 14.06$, $p = 0.00089$, 0 h vs. 12 h, $p = 0.0059$, 0 h vs. 24 h, $p = 0.0020$. (**C**,**D**) Pseudocolor plots of the D-value and the p value when auROC12 was compared with auROC0 (**C**) or when auROC24 was compared with auROC0 (**D**). (**E**) Cumulative probability of auROCs. Two-sample *K-S* test: 0 h vs. 12 h, $p = 0.0092$, 0 h vs. 24 h, $p = 0.0024$, 12 h vs. 24 h, $p = 0.90$. ** $p < 0.01$.

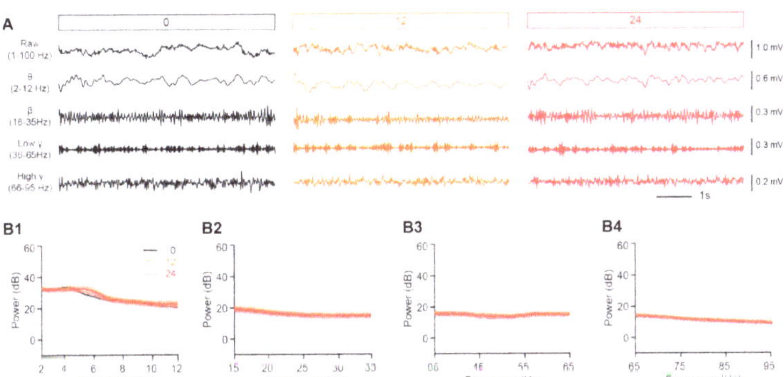

Figure 4. Nutritional status has no significant effect on the ongoing LFP in the OB. (**A**) Examples of ongoing baseline LFP signals from a single mouse fasted for 0 h (black), 12 h (orange), and 24 h (red). The first row shows 6 s of the raw trace; the second to fifth rows show the filtered signal (theta, beta, low gamma, and high gamma, respectively). (**B**) The averaged power spectrum of the ongoing LFP signals. (**B1**–**B4**) show the averaged power spectrum in the theta (**B1**), beta (**B2**), low gamma (**B3**), and high gamma (**B4**) bands across the group of mice. (**B1**). Friedman's test: χ^2 (2,38) = 0.57, $p = 0.75$. (**B2**). Friedman's test: χ^2 (2, 80) = 2.06, $p = 0.36$. (**B3**). Friedman's test: χ^2 (2, 118) = 2.85, $p = 0.24$. (**B4**). One-way rANOVA: F (2,116) = 1.09, $p = 0.34$. Error bars show the SEM.

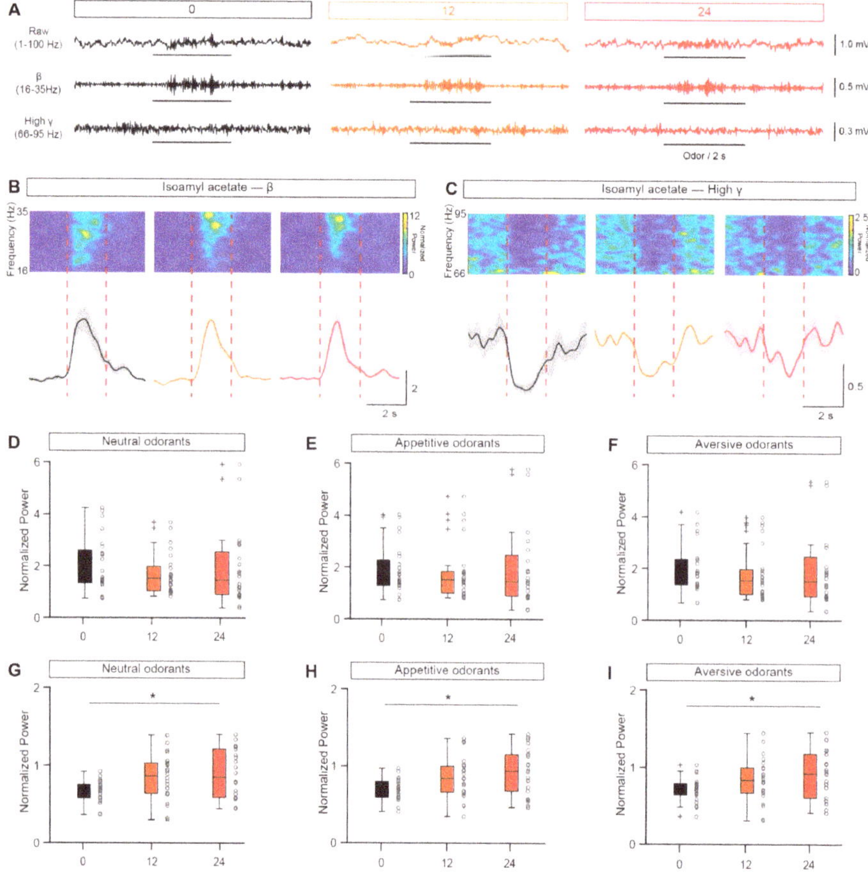

Figure 5. Nutritional status modulates odor-evoked LFP responses. (**A**) Responses of the raw LFP trace and the filtered beta and high gamma bands to odor stimulation under different nutritional states. Black bars indicate odor stimulation. (**B**,**C**) Top: Example power spectra for odor-evoked beta (**B**) and high gamma oscillations in the OB when mice were fasted for 0 h (black), 12 h (orange), or 24 h (red). Bottom: Trial-averaged normalized traces of odor-evoked beta (**B**) and high gamma (**C**) responses. The red dotted lines indicate the period of odor stimulation. Error bars show the SEM. (**D**–**F**) Comparison of the power in the normalized odor-evoked beta band evoked by neutral (**D**), appetitive (**E**), or aversive odorants (**F**) under different fasting states (n = 12, neutral odorants, Friedman's test: $\chi^2_{(2,46)} = 1$, $p = 0.61$; appetitive odorants, Friedman's test: $\chi^2_{(2,46)} = 0.58$, $p = 0.75$; aversive odorants, Friedman's test: $\chi^2_{(2,46)} = 4.08$, $p = 0.13$). (**G**–**I**) Comparison of the power in the normalized odor-evoked high gamma evoked by neutral (**G**), appetitive (**H**), or aversive odorants (**I**) under different fasting conditions (n = 12, neutral odorants, Friedman's test: $\chi^2_{(2,42)} = 8.27$, $p = 0.016$, 0 h vs. 24 h, $p = 0.012$; appetitive odorants, Friedman's test: $\chi^2_{(2,42)} = 7.64$, $p = 0.022$, 0 h vs. 24 h, $p = 0.018$; aversive odorants, Friedman's test: $\chi^2_{(2,42)} = 8.45$, $p = 0.015$, 0 h vs. 24 h, $p = 0.018$). * $p < 0.05$.

When the odorant-induced changes in LFP power under satiated and fasted states were surveyed across all mice and odors (Figure 5D–I), we found that there was no significant difference between satiated and fasted states for the odor-evoked beta response (Figure 5D–F), but the high gamma response was significantly reduced for all the three types of odorants with the longest fasting duration

(48 h after the removal of food, Figure 5G–I). These results indicate that odor-evoked inhibition of the neural network in the OB is reduced in a fasted state in awake, head-fixed mice.

3.4. Slight Decrease in the Sniffing Volume Between Satiated and Fasted States

Since mice rely on respiration/sniffing to sample odors, sniffing plays an important role in odor processing and representation in the OB [32]. Thus, we next investigated whether sniffing changes under satiated and fasted states. The raw sniffing patterns recorded under different states are shown in Figure 6A. We analyzed different aspects of the sniffing signal, including sniffing frequency, mean inhalation duration (MID), and volume (Figure 6B). As shown in Figure 6C, there were no significant changes in ongoing sniffing frequency, MID, or volume in the fasted state versus the satiated state. Analysis of the odor-evoked sniffing data across all animals supports the result that only sniffing volume changed with fasting (24 h after the removal of food); this phenomenon was observed consistently for all three types of odorant (Figure 6D–F). Thus, taken together, these results indicate that there is only a weak difference in the sniffing pattern between satiated and fasted states, limited to a slight decrease in sniffing volume.

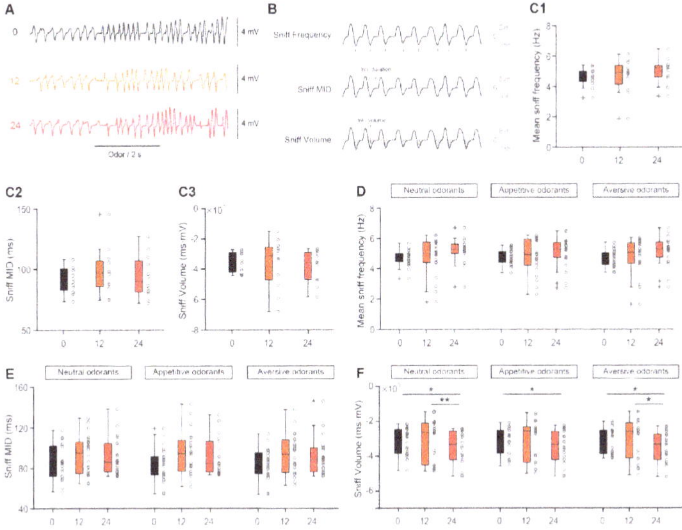

Figure 6. Change in the sniffing volume under different nutritional states. (**A**) Raw sniff trace recorded from a representative mouse fasted for 0 h (black), 12 h (orange), or 24 h (red). (**B**) Diagram illustrating the extraction of sniff frequency, mean inhalation duration (MID), and volume from a sample nasal flow trace. (**C**) Sniffing recorded under baseline conditions (no odor presentation) after fasting for different durations. (**C1**). Mean sniff frequency. One-way ANOVA: $F_{(2,22)} = 1.30$, $p = 0.29$. (**C2**). Sniff mean inhalation duration (MID). Friedman's test: $\chi^2_{(2,22)} = 2.17$, $p = 0.34$. (**C3**). Sniff volume. Friedman's test: $\chi^2_{(2,22)} = 3.17$, $p = 0.21$. (**D**) Odor-evoked mean sniff frequency recorded when mice were fasted for 0 h (black), 12 h (orange), or 24 h (red). Neutral odorants: Friedman's test: $\chi^2_{(2,42)} = 3.91$, $p = 0.14$. Appetitive odorants: Friedman's test: $\chi^2_{(2,42)} = 3.27$, $p = 0.19$. Aversive odorants: Friedman's test: $\chi^2_{(2,42)} = 4.73$, $p = 0.094$. (**E**) Fasting has no effect on odor-evoked sniff MID. Neutral odorants: Friedman's test: $\chi^2_{(2,42)} = 1.91$, $p = 0.38$. Appetitive odorants: Friedman's test: $\chi^2_{(2,42)} = 2.45$, $p = 0.29$. Aversive odorants: Friedman's test: $\chi^2_{(2,42)} = 1.91$, $p = 0.38$. (**F**) Odor-evoked sniff volume changes with fasting. Neutral odorants: Friedman's test: $\chi^2_{(2,42)} = 10.18$, $p = 0.0062$, 0 h vs. 24 h, $p = 0.042$, 12 h vs. 24 h, $p = 0.0072$. Appetitive odorants: Friedman's test: $\chi^2_{(2,42)} = 6.91$, $p = 0.032$, 0 h vs. 24 h, $p = 0.042$. Aversive odorants: Friedman's test: $\chi^2_{(2,42)} = 9.91$, $p = 0.0071$, 0 h vs. 24 h, $p = 0.028$, 12 h vs. 24 h, $p = 0.012$. * $p < 0.05$, ** $p < 0.01$.

4. Discussion

In this study, we investigated how nutritional states modulate the neural activity and neural representation of odors in the M/Ts of awake, head-fixed mice. Our data indicate that, in a fasted state, both the spontaneous firing rate of M/Ts and the number of cells showing an excitatory response to odors are increased at the single-cell level, suggesting that the excitability of neural activity is enhanced. Unexpectedly, the ability of M/Ts to discriminate different odors is slightly decreased in the fasted state, indicating that the effect of different nutritional states on olfaction is complex and that higher brain centers beyond the OB are likely involved.

Many previous studies have investigated behavioral odor detection and discrimination under different nutritional states in humans and rodents [2,5]. Fasting results in an increased perception of some food-related odors in humans [33] and olfactory sensitivity to a neutral odor increases in fasted rats [7]. The neural mechanisms underlying these behavioral observations are not clear, even though many studies have focused on this issue and a lot of data have been collected [16,17,21,34–37]. Since the OB is the first olfactory center and plays critical roles in odor information processing, most of the studies of different nutritional states examined neural activity in the OB [2,16–18,34,37]. In the 1970s, Pager found that the electrical activity of mitral cells was dependent on nutritional status: fasting selectively increased mitral cell multiunit responses to food odor [16]. This finding was partially supported by another study in which mitral cell single-unit responses to odor increased with fasting, regardless of the odorant [17]. Our results are consistent with the latter since we found that both the number of M/Ts showing an excitatory response and the amplitude of the response were greater in a fasted state, and that these increments were independent of odor type. In addition, we found that the baseline firing rate of single M/Ts was significantly increased in a fasted state. Therefore, the excitability of M/T neural activity is enhanced in the OB of fasted mice.

While the spikes from single unit in the OB have the capacity to encode odor identity in awake behaving rodents [13,22], LFP recorded form the OB reflect temporally coordinated neuronal ensembles and provides a reference for spike timing-based codes [38]. In the OB, the spikes from M/Ts are highly correlated with high-gamma oscillation and their correlation carries important information on odor identity [22,39]. Thus, it is not surprising that the increase in excitatory neural activity in the OB in a fasted state is further supported by our LFP recordings. Although the ongoing baseline LFP activity did not change from the satiated to the fasted state, the amplitude of the odor-evoked gamma oscillations was significantly decreased, indicating an increase in an excitatory component of the LFP or a decrease in an inhibitory component during odor stimulation in a fasted state. Since gamma oscillations in the OB are generated by the an interaction between M/Ts and granule cells [39,40], and since the firing rate of M/Ts is greater in the fasted state, the decrease in the odor response in the fasted state is likely due to the increased excitability of the M/Ts. Whether the firing of granule cells changes with fasting is an interesting question; cell-type-specific recording of the spikes from granule cells will be needed to address this, using, for example, the juxtacellular "loose-patch" recording method [41].

Although previous study performed in free-moving rats found that odor-evoked beta response was affected by fasting [34], we did not find significant changes after fasting during passive odor stimulation in present study. However, it is likely that fasting would change odor-evoked beta oscillation if the mice detect the odors actively, e.g., during the odor discrimination task, since beta oscillation is critically involved in the learning process [39,42]. Previous studies have demonstrated that beta oscillation is generated by the interaction between the OB and piriform cortex [43], and the centrifugal input from the piriform cortex to the OB should be critical for the beta oscillation. Strikingly, CB1 receptors are expressed in these centrifugal fibers and they play important roles in food intake [1]. Thus, odor-evoked beta oscillation is likely modulated by fasting during active rather than passive odor sampling.

Another important finding in our study is that the neural discrimination of odors by M/Ts is slightly decreased in the fasted state. Interestingly, for individual odor pairs, a significant decrease in odor discrimination was found only for food odor and 2,4,5-trimethylthiazole (Figure 3). Since behaviorally, odor discrimination is better during a fasted state [7,33], this finding was unexpected. One possible explanation is that the mice in our experiment were head-fixed when the odor was delivered rather than free-moving. The experiment was performed with head fixation since odor sampling is more stable with head fixation and more dynamic when mice are free moving [12]. Although the behavioral output of the animals is generally similar under the two conditions [44], the neural activity may be very different because of the different status. Another possible explanation for the unexpected decrease in neural discrimination of odors is that the mice in our study may have been over-fasted, since we observed that the sniffing volume decreased slightly under a fasted state. Thus, the decrease in odor discrimination by single M/Ts in a fasted state is likely due to changes in the animals' general state, such as the metabolic condition. Furthermore, we used the firing rate to calculate odor representation in the present study; however, temporal information can also be used for odor representation in the OB [13,14,45]. Finally, the OB receives dense modulation from higher brain areas, including both feedback and centrifugal inputs, and all of these projections dramatically modulate cell activity, odor information processing, and odor representation in the OB [12,46–48]. The change in odor discrimination at the single cell level observed in the OB is the result of complex network interactions between the OB and other higher brain centers. Thus, the neural representation in the OB may not necessarily represent the final behavioral output.

In the OB, activity in the circuits and in the different neuronal subtypes is modulated dramatically by sniffing [32]. The changes in M/T neural activity and odor discrimination in the fasted state observed in the present study may be due to changes in the sniffing pattern. However, we found only a slight decrease in sniffing volume with fasting; both sniffing frequency and MID remained unchanged. This indicates that the modulation of neural activity in the OB by nutritional state is independent of sniffing. Interestingly, a higher sniffing frequency has been reported in fasted free-moving rats [49], but we did not observe a significant change in sniffing frequency in head-fixed mice. This discrepancy is likely due to differences between the free-moving and head-fixed states. In free-moving state, animals are under active condition and they can move around and locate potential food source by fast sniffing. However, in head-fixed state, animals are under passive condition, cannot not move, and thus have no enthusiasms to find food and need not sniff fast. Thus, this discrepancy between the two studies is likely due to differences between the free-moving and head-fixed states. Future studies could test this possibility by monitoring sniffing signals in the same animals under free-moving and head-fixed conditions.

In summary, we found that the excitability of neural activity is enhanced but neural discrimination of odors is slightly decreased in the OB of fasted mice under awake, head-fixed conditions. Although a more detailed investigation into the underlying neural mechanisms is warranted, our results represent a first step toward understanding the neural circuit mechanisms by which olfaction is modulated by nutritional status.

Author Contributions: Data curation, J.W., P.L., F.C., L.G., and Y.L.; Formal analysis, J.W., P.L., L.G., Y.L., and A.L.; Funding acquisition, F.C., and A.L.; Writing—original draft, J.W., P.L., F.C., L.G., Y.L., and A.L.; Writing—review & editing, J.W., P.L., F.C., L.G., Y.L., and A.L. All authors have read and agreed to the published version of the manuscript.

Funding: This work was funded by the National Natural Science Foundation of China (NSFC, 31571082 and 31872771 to A.L), the Priority Academic Program Development of Jiangsu Higher Education Institutions (16KJA180007 to A.L), and the National Science Foundation of Jiangsu Higher Education Institutions of China (18KJD310006 to F.C).

Acknowledgments: We thank Han Xu and Xinsong Guo for technical assistance and Qing Liu for critical comments.

Conflicts of Interest: The authors declare no conflict of interest.

References

1. Soria-Gomez, E.; Bellocchio, L.; Reguero, L.; Lepousez, G.; Martin, C.; Bendahmane, M.; Ruehle, S.; Remmers, F.; Desprez, T.; Matias, I.; et al. The endocannabinoid system controls food intake via olfactory processes. *Nat. Neurosci.* **2014**, *17*, 407–415. [CrossRef]
2. Palouzier-Paulignan, B.; Lacroix, M.C.; Aime, P.; Baly, C.; Caillol, M.; Congar, P.; Julliard, A.K.; Tucker, K.; Fadool, D.A. Olfaction under metabolic influences. *Chem. Senses* **2012**, *37*, 769–797. [CrossRef]
3. Rolls, E.T. Taste, olfactory, and food texture processing in the brain, and the control of food intake. *Physiol. Behav.* **2005**, *85*, 45–56. [CrossRef]
4. Yeomans, M.R. Olfactory influences on appetite and satiety in humans. *Physiol. Behav.* **2006**, *89*, 10–14. [CrossRef]
5. Julliard, A.K.; Al Koborssy, D.; Fadool, D.A.; Palouzier-Paulignan, B. Nutrient Sensing: Another Chemosensitivity of the Olfactory System. *Front. Physiol.* **2017**, *8*. [CrossRef]
6. Julliard, A.K.; Chaput, M.A.; Apelbaum, A.; Aime, P.; Mahfouz, M.; Duchamp-Viret, P. Changes in rat olfactory detection performance induced by orexin and leptin mimicking fasting and satiation. *Behav. Brain Res.* **2007**, *183*, 123–129. [CrossRef]
7. Aime, P.; Duchamp-Viret, P.; Chaput, M.A.; Savigner, A.; Mahfouz, M.; Julliard, A.K. Fasting increases and satiation decreases olfactory detection for a neutral odor in rats. *Behav. Brain Res.* **2007**, *179*, 258–264. [CrossRef]
8. Albrecht, J.; Schreder, T.; Kleemann, A.M.; Schopf, V.; Kopietz, R.; Anzinger, A.; Demmel, M.; Linn, J.; Kettenmann, B.; Wiesmann, M. Olfactory detection thresholds and pleasantness of a food-related and a non-food odour in hunger and satiety. *Rhinology* **2009**, *47*, 160–165.
9. McClintock, T.S. Achieving singularity in mammalian odorant receptor gene choice. *Chem. Senses* **2010**, *35*, 447–457. [CrossRef]
10. Li, Q.; Tachie-Baffour, Y.; Liu, Z.; Baldwin, M.W.; Kruse, A.C.; Liberles, S.D. Non-classical amine recognition evolved in a large clade of olfactory receptors. *eLife* **2015**, *4*, e10441. [CrossRef]
11. Buck, L.; Axel, R. A novel multigene family may encode odorant receptors: A molecular basis for odor recognition. *Cell* **1991**, *65*, 175–187. [CrossRef]
12. Li, A.; Rao, X.; Zhou, Y.; Restrepo, D. Complex neural representation of odour information in the olfactory bulb. *Acta Physiol.* **2020**, *228*, e13333. [CrossRef]
13. Uchida, N.; Poo, C.; Haddad, R. Coding and transformations in the olfactory system. *Annu. Rev. Neurosci.* **2014**, *37*, 363–385. [CrossRef]
14. Chong, E.; Rinberg, D. Behavioral readout of spatio-temporal codes in olfaction. *Curr. Opin. Neurobiol.* **2018**, *52*, 18–24. [CrossRef]
15. Hu, B.; Geng, C.; Hou, X.Y. Oligomeric amyloid-beta peptide disrupts olfactory information output by impairment of local inhibitory circuits in rat olfactory bulb. *Neurobiol. Aging* **2017**, *51*, 113–121. [CrossRef] [PubMed]
16. Pager, J.; Giachetti, I.; Holley, A.; Le Magnen, J. A selective control of olfactory bulb electrical activity in relation to food deprivation and satiety in rats. *Physiol. Behav.* **1972**, *9*, 573–579. [CrossRef]
17. Apelbaum, A.F.; Chaput, M.A. Rats habituated to chronic feeding restriction show a smaller increase in olfactory bulb reactivity compared to newly fasted rats. *Chem. Senses* **2003**, *28*, 389–395. [CrossRef]
18. Sun, C.; Tang, K.; Wu, J.; Xu, H.; Zhang, W.; Cao, T.; Zhou, Y.; Yu, T.; Li, A. Leptin modulates olfactory discrimination and neural activity in the olfactory bulb. *Acta Physiol.* **2019**, *227*, e13319. [CrossRef]
19. Wang, D.; Liu, P.; Mao, X.; Zhou, Z.; Cao, T.; Xu, J.; Sun, C.; Li, A. Task-Demand-Dependent Neural Representation of Odor Information in the Olfactory Bulb and Posterior Piriform Cortex. *J. Neurosci.* **2019**, *39*, 10002–10018. [CrossRef]
20. Zhang, W.; Sun, C.; Shao, Y.; Zhou, Z.; Hou, Y.; Li, A. Partial depletion of dopaminergic neurons in the substantia nigra impairs olfaction and alters neural activity in the olfactory bulb. *Sci. Rep.* **2019**, *9*, 254. [CrossRef]
21. Zhou, Y.; Wang, X.; Cao, T.; Xu, J.; Wang, D.; Restrepo, D.; Li, A. Insulin Modulates Neural Activity of Pyramidal Neurons in the Anterior Piriform Cortex. *Front. Cell Neurosci.* **2017**, *11*, 378. [CrossRef]
22. Li, A.; Gire, D.H.; Restrepo, D. Upsilon spike-field coherence in a population of olfactory bulb neurons differentiates between odors irrespective of associated outcome. *J. Neurosci.* **2015**, *35*, 5808–5822. [CrossRef]

23. Li, A.; Guthman, E.M.; Doucette, W.T.; Restrepo, D. Behavioral status influences the dependence of odorant-induced change in firing on pre-stimulus firing rate. *J. Neurosci.* **2017**. [CrossRef]
24. Kobayakawa, K.; Kobayakawa, R.; Matsumoto, H.; Oka, Y.; Imai, T.; Ikawa, M.; Okabe, M.; Ikeda, T.; Itohara, S.; Kikusui, T.; et al. Innate versus learned odour processing in the mouse olfactory bulb. *Nature* **2007**, *450*, 503–508. [CrossRef]
25. Muthusamy, N.; Zhang, X.; Johnson, C.A.; Yadav, P.N.; Ghashghaei, H.T. Developmentally defined forebrain circuits regulate appetitive and aversive olfactory learning. *Nat. Neurosci.* **2017**, *20*, 20–23. [CrossRef]
26. Iurilli, G.; Datta, S.R. Population Coding in an Innately Relevant Olfactory Area. *Neuron* **2017**, *93*, 1180–1197. [CrossRef]
27. Saraiva, L.R.; Kondoh, K.; Ye, X.; Yoon, K.H.; Hernandez, M.; Buck, L.B. Combinatorial effects of odorants on mouse behavior. *Proc. Natl. Acad. Sci. USA* **2016**, *113*, E3300–E3306. [CrossRef]
28. Fawcett, T. An introduction to ROC analysis. *Pattern Recogn. Lett.* **2006**, *27*, 861–874. [CrossRef]
29. Buzsaki, G.; Anastassiou, C.A.; Koch, C. The origin of extracellular fields and currents–EEG, ECoG, LFP and spikes. *Nat. Rev. Neurosci.* **2012**, *13*, 407–420. [CrossRef]
30. Lowry, C.A.; Kay, L.M. Chemical factors determine olfactory system beta oscillations in waking rats. *J. Neurophysiol.* **2007**, *98*, 394–404. [CrossRef]
31. Beshel, J.; Kopell, N.; Kay, L.M. Olfactory bulb gamma oscillations are enhanced with task demands. *J. Neurosci.* **2007**, *27*, 8358–8365. [CrossRef]
32. Wachowiak, M. All in a Sniff: Olfaction as a Model for Active Sensing. *Neuron* **2011**, *71*, 962–973. [CrossRef]
33. Mulligan, C.; Moreau, K.; Brandolini, M.; Livingstone, B.; Beaufrere, B.; Boirie, Y. Alterations of sensory perceptions in healthy elderly subjects during fasting and refeeding. A pilot study. *Gerontology* **2002**, *48*, 39–43. [CrossRef]
34. Chabaud, P.; Ravel, N.; Wilson, D.A.; Mouly, A.M.; Vigouroux, M.; Farget, V.; Gervais, R. Exposure to behaviourally relevant odour reveals differential characteristics in rat central olfactory pathways as studied through oscillatory activities. *Chem. Senses* **2000**, *25*, 561–573. [CrossRef]
35. Pager, J. Ascending olfactory information and centrifugal influxes contributing to a nutritional modulation of the rat mitral cell responses. *Brain Res.* **1978**, *140*, 251–269. [CrossRef]
36. Al Koborssy, D.; Palouzier-Paulignan, B.; Canova, V.; Thevenet, M.; Fadool, D.A.; Julliard, A.K. Modulation of olfactory-driven behavior by metabolic signals: Role of the piriform cortex. *Brain Struct. Funct.* **2018**. [CrossRef]
37. Chelminski, Y.; Magnan, C.; Luquet, S.H.; Everard, A.; Meunier, N.; Gurden, H.; Martin, C. Odor-Induced Neuronal Rhythms in the Olfactory Bulb Are Profoundly Modified in ob/ob Obese Mice. *Front. Physiol.* **2017**, *8*, 2. [CrossRef]
38. Losacco, J.; Ramirez-Gordillo, D.; Gilmer, J.; Restrepo, D. Learning improves decoding of odor identity with phase-referenced oscillations in the olfactory bulb. *eLife* **2020**, *9*. [CrossRef]
39. Kay, L.M.; Beshel, J.; Brea, J.; Martin, C.; Rojas-Libano, D.; Kopell, N. Olfactory oscillations: The what, how and what for. *Trends Neurosci.* **2009**, *32*, 207–214. [CrossRef]
40. Kay, L.M. Olfactory system oscillations across phyla. *Curr. Opin. Neurobiol.* **2015**, *31*, 141–147. [CrossRef]
41. Cazakoff, B.N.; Lau, B.Y.; Crump, K.L.; Demmer, H.S.; Shea, S.D. Broadly tuned and respiration-independent inhibition in the olfactory bulb of awake mice. *Nat. Neurosci.* **2014**, *17*, 569–576. [CrossRef]
42. Martin, C.; Ravel, N. Beta and gamma oscillatory activities associated with olfactory memory tasks: Different rhythms for different functional networks? *Front. Behav. Neurosci.* **2014**, *8*, 218. [CrossRef]
43. Neville, K.R.; Haberly, L.B. Beta and gamma oscillations in the olfactory system of the urethane-anesthetized rat. *J. Neurophysiol.* **2003**, *90*, 3921–3930. [CrossRef]
44. Abraham, N.M.; Guerin, D.; Bhaukaurally, K.; Carleton, A. Similar odor discrimination behavior in head-restrained and freely moving mice. *PLoS ONE* **2012**, *7*, e51789. [CrossRef]
45. Cury, K.M.; Uchida, N. Robust odor coding via inhalation-coupled transient activity in the mammalian olfactory bulb. *Neuron* **2010**, *68*, 570–585. [CrossRef]
46. Padmanabhan, K.; Osakada, F.; Tarabrina, A.; Kizer, E.; Callaway, E.M.; Gage, F.H.; Sejnowski, T.J. Diverse Representations of Olfactory Information in Centrifugal Feedback Projections. *J. Neurosci.* **2016**, *36*, 7535–7545. [CrossRef]
47. Lizbinski, K.M.; Dacks, A.M. Intrinsic and Extrinsic Neuromodulation of Olfactory Processing. *Front. Cell Neurosci.* **2017**, *11*, 424. [CrossRef]

48. Fletcher, M.L.; Chen, W.R. Neural correlates of olfactory learning: Critical role of centrifugal neuromodulation. *Learn. Mem.* **2010**, *17*, 561–570. [CrossRef]
49. Prud'homme, M.J.; Lacroix, M.C.; Badonnel, K.; Gougis, S.; Baly, C.; Salesse, R.; Caillol, M. Nutritional status modulates behavioural and olfactory bulb Fos responses to isoamyl acetate or food odour in rats: Roles of orexins and leptin. *Neuroscience* **2009**, *162*, 1287–1298. [CrossRef]

Article

Modulation of Sex Pheromone Discrimination by a UDP-Glycosyltransferase in *Drosophila melanogaster*

Stéphane Fraichard [1,†], Arièle Legendre [1,†], Philippe Lucas [2], Isabelle Chauvel [1], Philippe Faure [1], Fabrice Neiers [1], Yves Artur [1], Loïc Briand [1], Jean-François Ferveur [1] and Jean-Marie Heydel [1,*]

1 Centre des Sciences du Goût et de l'Alimentation, AgroSup Dijon, CNRS, INRAE, Université Bourgogne Franche -Comté, F-21000 Dijon, France; stephane.fraichard@u-bourgogne.fr (S.F.); ariele_l@hotmail.com (A.L.); isabelle.chauvel@u-bourgogne.fr (I.C.); philippe.faure@u-bourgogne.fr (P.F.); fabrice.neiers@u-bourgogne.fr (F.N.); yves.artur@u-bourgogne.fr (Y.A.); loic.briand@inra.fr (L.B.); jean-francois.ferveur@u-bourgogne.fr (J.-F.F.)

2 Institut d'Ecologie et des Sciences de l'Environnement de Paris, F-78000 Versailles, France; philippe.lucas@inra.fr

* Correspondence: jean-marie.heydel@u-bourgogne.fr

† These authors contributed equally to this work.

Received: 27 January 2020; Accepted: 21 February 2020; Published: 25 February 2020

Abstract: The detection and processing of chemical stimuli involve coordinated neuronal networks that process sensory information. This allows animals, such as the model species *Drosophila melanogaster*, to detect food sources and to choose a potential mate. In peripheral olfactory tissues, several classes of proteins are acting to modulate the detection of chemosensory signals. This includes odorant-binding proteins together with odorant-degrading enzymes (ODEs). These enzymes, which primarily act to eliminate toxic compounds from the whole organism also modulate chemodetection. ODEs are thought to neutralize the stimulus molecule concurrently to its detection, avoiding receptor saturation thus allowing chemosensory neurons to respond to the next stimulus. Here, we show that one UDP-glycosyltransferase (UGT36E1) expressed in *D. melanogaster* antennal olfactory sensory neurons (OSNs) is involved in sex pheromone discrimination. UGT36E1 overexpression caused by an insertion mutation affected male behavioral ability to discriminate sex pheromones while it increased OSN electrophysiological activity to male pheromones. Reciprocally, the decreased expression of UGT36E1, controlled by an RNAi transgene, improved male ability to discriminate sex pheromones whereas it decreased electrophysiological activity in the relevant OSNs. When we combined the two genotypes (mutation and RNAi), we restored wild-type-like levels both for the behavioral discrimination and UGT36E1 expression. Taken together, our results strongly suggest that this UGT plays a pivotal role in Drosophila pheromonal detection.

Keywords: olfaction; chemosensory; odorant-degrading enzymes; perireceptor; drosophila

1. Introduction

The initiation of olfaction is mediated by olfactory receptors which interact with a variety of odorant molecules allowing their fine detection and discrimination. Receptor activation triggers the olfactory signal transmitted as a depolarization train of spikes to the central nervous system, this resulting in its integrated perception and an adapted behavioral response. The detection of odorants is modulated by perireceptor events through which a complex series of biochemical processes are carried out to influence the entry, exit and/or residence time of odorant molecules in the receptor environment [1,2]. Perireceptor mechanisms play a significant role in the modulation of the stimulus availability for receptors, therefore impacting its reception and subsequent perception. They include, first, the binding and transport of hydrophobic odorant molecules by odorant-binding proteins (OBPs) [3–5] and then their inactivation and elimination by detoxification enzymes called odorant-metabolizing enzymes in vertebrate (OMEs) [6] or odorant-degrading enzymes (ODEs) in insect [2,7]. Indeed, some metabolizing enzymes can be expressed, either specifically, and/or at a high concentration, in the olfactory tissues or organs [6,8–11]. These enzymes are primarily involved in detoxification processes by catalyzing the biotransformation of hydrophobic xenobiotic molecules through two phases often but not necessarily successive. During phase I, functionalization enzymes introduce, or unmask, functional groups into xenobiotics through oxidation, reduction or hydrolysis reaction (e.g., cytochrome P450, CYP; aldehyde dehydrogenase; carboxylesterase, CES). During phase II, functionalized metabolites can be conjugated to hydrophilic products by transferases enzymes such as UDP-glycosyltransferases (UGT) or glutathione transferases (GST). Thus, the resulting inactive hydrophilic metabolites can be easily eliminated. These enzymes are organized in networks allowing to metabolize a broad range of substrates.

In insects, the characterization of metabolizing enzymes received an increasing interest with regard to their role in insecticide resistance, adaptation to host plant volatile and their function; as ODEs, in the termination of the olfactory signal to maintain a relatively high olfactory sensitivity toward new stimuli. In particular, recent studies investigating the antennal transcriptome in different species identified varied ODEs including CYP, CES, GST and UGT [11–24]. These reports have completed and confirmed the case-by-case identification of previously characterized ODEs [25–31]. Altogether, a high number of diverse antennal ODEs have been identified including among others, 30 CES and 84 CYP in *Spodoptera littoralis* [14,26,27,32,33], and 31 GST and 57 CYP in *Drosophila melanogaster* antennae [11].

However, data about ODE function in chemosensory process are still limited, likely because of (i) the current focus on odorant/receptor interaction and (ii) the complexity due to the high diversity of enzymes and odorant substrates in different species and strains. Phase I ODE's function was initially investigated with regard to the perception of pheromones [2,34,35] because of their critical role and their ability to trigger specific and measurable sexual behavior [7,36]. Although phase II enzymes are expected to play a similar role as phase I enzymes in the metabolism of plant volatiles, pheromones or diverse odorants [9,16,20,22], no olfactory function was revealed so far in insects. Among phase II ODEs, the study of UGTs received an increasing interest during the last decade, especially with regard to insecticide resistance and plant defense mechanisms [37–44]. This major class of enzymes in the animal kingdom [45,46] catalyzes in insects the conjugation of a glycosyl group brought by a UDP-glycoside to hydrophobic substrates [47,48]. The fact that UGT expression is enriched in antennae supports its potential role in olfaction [11,47,49–51], in relation with its high number of isoforms: 20 in *Holotrichia parallela Motschulsky* and 11 in *S. littoralis* and 19 in *D. melanogaster* [11].

In the present study, we combined molecular, genetic, behavioral and electrophysiological approaches to investigate the influence of a phase II ODE, a UGT (UGT36E1), on the ability of *D. melanogaster* males to discriminate sex pheromones. UGT36E1 (CG17322) is one of the most expressed UGT in *D. melanogaster* [47]. We used both mutational and interferential RNA approaches to target UGT36E1 expression in the fly olfactory tissues. We found that both the decreased and increased UGT expression in the peripheral olfactory system affected sex pheromone discrimination in a reciprocal manner.

2. Materials and Methods

2.1. Stocks and Flies

All *D. melanogaster* strains were raised on yeast/cornmeal/agar medium and kept at 24 ± 0.5 °C with 65 ± 5% humidity on a 12L:12D cycle. The wild-type Dijon2000 strain (Di2), used as a control, has been maintained in our lab for more than a decade. The P-UGT36E1 mutant strain (y1 w67c23); P{SUPor-P}CG17322KG04070 (#13518) and the transgenic strains *neur*-GAL4 (#6393); *Orco*-GAL4

(#23292); UAS-CD8::GFP (#5130) were obtained from the Bloomington Drosophila Stock Center (Indiana University). The *Gr66a*-GAL4 strain was kindly provided by Dr. Hubert Amrein. RNAi transgenics against UGT36E1 were purchased from the VDRC [52]. To isogenize its genetic background, the P-UGT36E1 mutation was outcrossed to the Di2 wild-type genetic background of the w^{1118} strain by five successive backcrosses [53]. Crosses were performed using standard techniques and genetic methods [54].

2.2. Behavior

The behavior of single tester males towards two Di2 headless target flies (a male and a female) was measured over a 5 min period. All flies were isolated 0–4 h after eclosion under CO_2 anaesthesia. A few minutes prior to the test, anaesthetized target flies were decapitated with a razor blade (cleaned with ethanol between each sex). In all courtship tests, we used headless flies which do not copulate: therefore, decapitation allows courtship duration to be standardized. Moreover, some tests were performed under red light (25W with a Kodak safe-light filter no. 1) under which flies are virtually blind. In these conditions, most behavioral, visual and acoustic variables associated with the object fly are removed, this enhancing the behavioral effect of pheromones [55,56]. Only upright headless flies were used for the test. Tester males were individually aspirated (without anaesthesia) under a watch glass used as an observation chamber (1.6 cm^3). After 5 min, the two headless target flies were simultaneously introduced and the courtship index (CI) toward each target was measured (CIf = courtship towards female flies; CIm = courtship towards male flies). CI corresponds to the fraction of the time spent courting relative to the total amount of time multiplied by 100. We only took in account male active courtship (wing vibration, licking and attempting copulation; the very brief episodes of tapping behavior were rarely seen under red light and thus was not included in the CI calculation). No qualitative difference was noted between the different courtship sequences shown by the different tester male genotypes. The discrimination measure was based on the comparison between CIf and CIm for each genotype and condition used. $n > 30$. Locomotor activity, resulting of the cumulation of the activity of single flies, was measured during five periods of 10 s (total of 50 s), over five minutes ($n > 24$). All tests took place 1–5 h after lights on, in a room at 24 ± 0.5 °C with 65 ± 5% humidity. Tests were performed over several days to randomize the experimental variation of uncontrolled environmental parameters.

2.3. q-PCR

To process fly tissues, whole frozen flies were vortexed over two superposed meshes of different calibers. From bottom to top, tissue fractions successively consisted of the appendages (maxillary palps, antennal segments), heads (without appendages), and bodies (without heads or appendages). RNAs were extracted from *Drosophila*-fractionated tissues using RNAzol reagent (Euromedex, Souffelweyersheim, France) and treated with RNase-free DNase to avoid contamination by genomic DNA. Total RNA (1 µg) was reverse-transcribed using the iScript cDNA Synthesis Kit (BioRad, Hercules, USA). q-PCR reactions were carried out on a MyIQ (BioRad, Hercules, USA) using the IQ SYBR Green Supermix (BioRad, Hercules, USA). The q PCR conditions were as follows: 98 °C for 5 min to activate the hot-start DNA polymerase, followed by 40 cycles of 95 °C for 30 s, 65 °C for 30 s and 72 °C for 30 s. Each reaction was performed in triplicate and the mean of the three independent biological replicates (corresponding to three extractions) was calculated. All results were normalized to the RP49 and Actine5C mRNA level and calculated using the $\Delta\Delta$Ct method [57].

2.4. Expression Pattern of GAL4 Strains

The expression pattern of all GAL4 strains was always observed in F1 males resulting of the cross between UAS-GFP females and GAL4 males. F1 males were dissected and analyzed using fluorescent microscopy (Leica DM5000B, Leica Microsystems, Wetzlar, Germany). Frozen sections of antennae were collected and fixed in 4% formaldehyde in PBS for 30 min and washed with PBS. Antennal sections were then stained with goat anti-GFP (1:500; Rockland, Limerick, Ireland) and mouse anti-Elav (1:1000; Developmental Studies Hybridoma Bank). Detection of the different primary antibodies was carried out using AlexaFluor488 anti-goat (1:800, Molecular Probes, Eugene, USA) and AlexaFluor594 anti-mouse (1:800, Molecular Probes, Eugene, USA). After mounting in Vectashield (Vector Labs, Burlingame, CA, United States) images were made with a Leica TCS-SP2 confocal microscope.

2.5. Electrophysiology

Electroantennograms (EAGs) were recorded with Ringer-filled capillary glass electrodes (NaCl 120 mM; KCl 5 mM; $CaCl_2$ 1 mM; $MgCl_2$ 4 mM; HEPES 10 mM; pH 7.2) from 4- to 6-day-old adult male flies [58]. Immobilized flies were placed in a 1000 µL micropipet-tip, the extremity of the tip was filled with cotton and flies were blocked by dental wax. The ground electrode (tip diameter = 2–3 µm) was introduced into the right eye and the recording electrode (tip diameter = 10 µm) was placed on the distolateral surface of the third antennal segment. Recorded signals were amplified (×100) and low-pass filtered at 1 kHz using an Axopatch 200B amplifier (Molecular Devices, San José, USA) and monitored on a computer using a Digidata 1440A acquisition board (Molecular Devices, San José, USA) driven by Clampex 10 software (Molecular Devices, San José, USA).

Odorants were diluted 1/1000 (v/v) in paraffin oil and delivered from Pasteur pipettes containing 10 µL of diluted odorant deposited on a strip of filter paper. Stimulation with sex-specific odors was obtained by blowing a flow of humidified air (37.5 mL/s) for 1 s through a Pasteur pipette (inner diameter 5 mm, length 250 mm) containing 25 same-sex flies (4- to 6-day-old) [59]. Flies were placed in the tube one hour prior to experiments.

Flies were first stimulated with 2-Heptanone (2-H), then with odors of living males and finally with odors of living females. Flies were allowed to recover at least 2 min between each stimulation. The odorant 2-H molecule, which elicits a robust electrophysiological response, was chosen to normalize the EAG responses to male and female odors.

2.6. Statistical Analysis

For q-PCR, transcript level ratios were compared between strains (or body fractions) using Relative Expression Software Tool [57] (REST, REST-MCS beta software version 2) with 2000 iterations. This is based on the probability of an effect as large as that occurring under the null hypothesis (no effect of the treatment), using a randomization test (Pair-Wise Fixed Reallocation Randomization Test) [60]. For behavior and EAG, statistical analyses were performed using XLSTAT software. Within each genotype, the CI directed towards each sex target was compared with a Student's *t*-test. Since CIf and CIm are linked, they were compared between strains using ANOVA and LSD-Fischer post-hoc tests (sex target/strain genotype). Locomotor activity values were normally distributed and compared with a Mann-Whitney test ($p < 0.05$). The difference of EAG amplitude between stimulation was compared using a Wilcoxon bilateral test. EAG comparison between strains was performed with a Kruskal-Wallis test ($p < 0.05$).

3. Results/Discussion

We investigated in vivo the involvement of ODEs in the discrimination of sex pheromones in *D. melanogaster*. In this species, female and male pheromones diverge and induce reciprocal effects: they tend to stimulate or to inhibit male courtship behavior, respectively [61]. We discovered that a transposable P-element (P-UGT36E1) inserted into the UGT36E1 gene affected male discrimination of sex pheromones (Figure 1A). To determine the ability of single tester males to discriminate sex partners, we measured their courtship intensity (or courtship index = CI) toward both female and male target flies simultaneously presented [55]. This paradigm allowed us to measure their CI towards a female target fly (CIf) and towards a male (CIm). The CIf/CIm comparison allowed us to determine the ability of tester males to discriminate the two sex targets, for each genotype and experimental condition. We compared male discrimination ability under red light (Figure 1B; filled bars), in which visual stimuli provided by both target flies are ineffective and under white light (empty bars) allowing tester males to discriminate sex target based on their different morphology. Under red light, P-UGT36E1 homozygous mutant males showed similar CIm and CIf. In particular, mutant males showed a decreased CIf compared to wild-type tester males ($p = 0.0045$), while their CIm was not affected (p = ns). The loss of sex discrimination was likely due to a chemosensory defect given that mutant males tested under white light showed a strong preference to the female target, similarly to wild-type males (Figure 1B). This indicates that the mutation did not affect its overall sexual activity; moreover, the male ability to use visual cues to discriminate sexual partners is functional. Therefore, these results suggest that the P-UGT36E1 mutation affects the ability of male flies to discriminate sex pheromones.

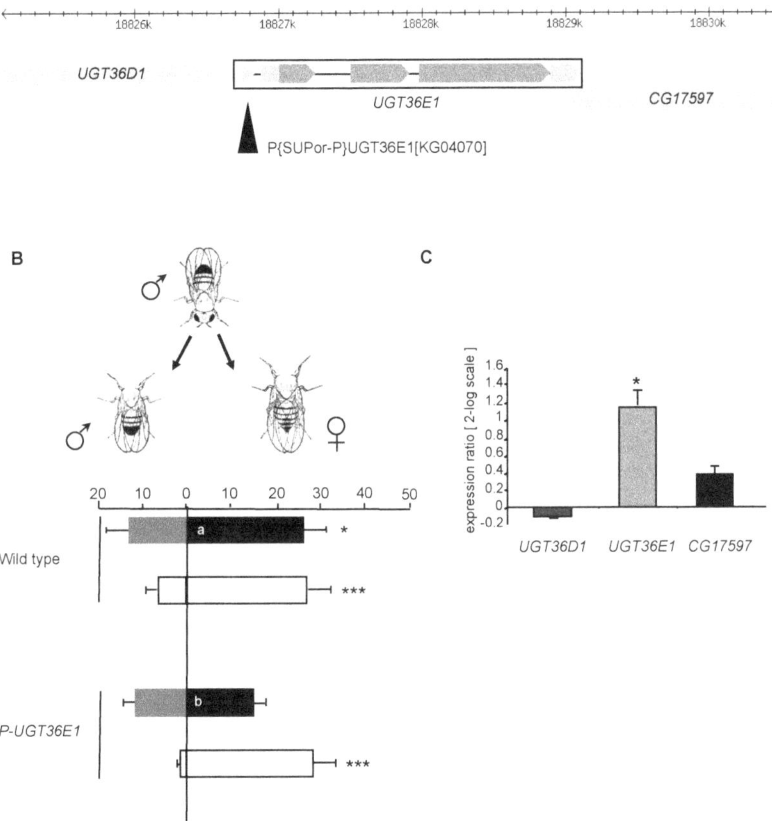

Figure 1. Effects of the P-element insertion in UGT36E1. (**A**) Schematic organization of the 37B1 chromosomal region. Arrowhead indicates the position of the P-element inserted in the 5′UTR region of UGT36E1. (**B**) The P-element inserted in UGT36E1 (P-UGT36E1) affects sex pheromone discrimination in male flies. Tests were carried out either under red light (filled bars) or white light (empty bars). Each mirrored bar represents the mean (± s.e.m.) courtship index towards female (CIf) and male (CIm). Individual 4-day-old tester males directed towards female (right) and male (left) headless targets, simultaneously presented during a 5 min observation period. Tester males were homozygous (P-UGT36E1) for the P-element mutation, or wild-type (Dijon strain); target flies always belonged to

this wild-type strain. Significant differences for male ability to discriminate between the two sexes are shown next to each mirrored bar as ***: $p < 0.001$; **: $p < 0.01$; *: $p < 0.05$ (Student's *t*-test). Courtship data towards each sex were tested using ANOVA and LSD Fisher tests (letters within bars indicates significant differences towards each sex). For each test, the number (n) was $n > 40$ (under red light) and $n > 30$ (under white light). **(C)** The P-element insertion affects RNA expression levels of UGT36E1. The expression levels of UGT36E1, UGT36D1 and CG17597 were analyzed by real-time PCR. The significance of differences in the ratio of transcript levels was based on a comparison between wild-type and P-UGT36E1 homozygous mutant flies (in log_2 scale control = 0). Data represent the mean (± s.e.m.) of the expression ratio (mutant: wild-type) carried out with three independent extractions.

Given that the P-element inserted upstream of the UGT36E1 gene is also in the vicinity of two other genes (UGT36D1, CG17597; Figure 1A), we measured mRNA expression of the three genes in P-UGT36E1 mutant males. Quantitative RT-PCR (q-PCR) revealed that only UGT36E1 significantly changed its mRNA expression (>two-fold increase in P-UGT36E1 flies as compared to controls; Figure 1C). The absence of any significant variation in expression between the head, thorax, abdomen, and appendages of either wild-type flies or mutant flies suggests that this UGT has not a tissue-specific expression (Figure S1). Moreover, this ubiquitous mutation-induced increase of UGT expression in P-UGT36E1 flies had no general behavioral effect given that both locomotor activity and global courtship index (CIf + CIm) were similar to those of wild-type flies under white light (Figure S2; Figure 1B).

To expand our investigation on the behavioral effect of UGT36E1, we used RNAi targeted against UGT36E1 (UAS-dsUGT36E1; hereafter dsUGT36E1) to knock down UGT36E1 RNA expression. We targeted dsUGT36E1 distinct subsets of tissues using several GAL4 drivers. We first used the *neuralized*-GAL4 (*neur*-GAL4) driver to target adult chemosensory organs involved in the perception and processing of pheromones (antenna, proboscis, wing margin and tarsa; Figure 2A–C). We found that *neur*-Gal4 is expressed in antennal neuronal cells (Figure 2D,E). Under red light, experimental males ("*neur*-GAL4/+, dsUGT36E1/+") showed increased sex discrimination compared to both control parental transgenic genotypes (*neur*-GAL4/+ and dsUGT36E1/+; Figure 2F). This effect was mostly due to the lower CIm shown by knockdown flies (Cim = 6.5 vs. 16.5 for dsUGT36E1/+ control males; $p = 0.022$). However, CIf were similar in mutant and control tester males (CIf = 28 and 32, respectively). Moreover, *neur*-GAL4/+, dsUGT36E1/+ males showed a 6-fold decrease for mRNA expression level in sensory appendages and heads compared to dsUGT36E1/+ controls ($p = 0.031$; Figure 2G). Differently, no expression difference was detected between the abdomen and thorax of the two male genotypes.

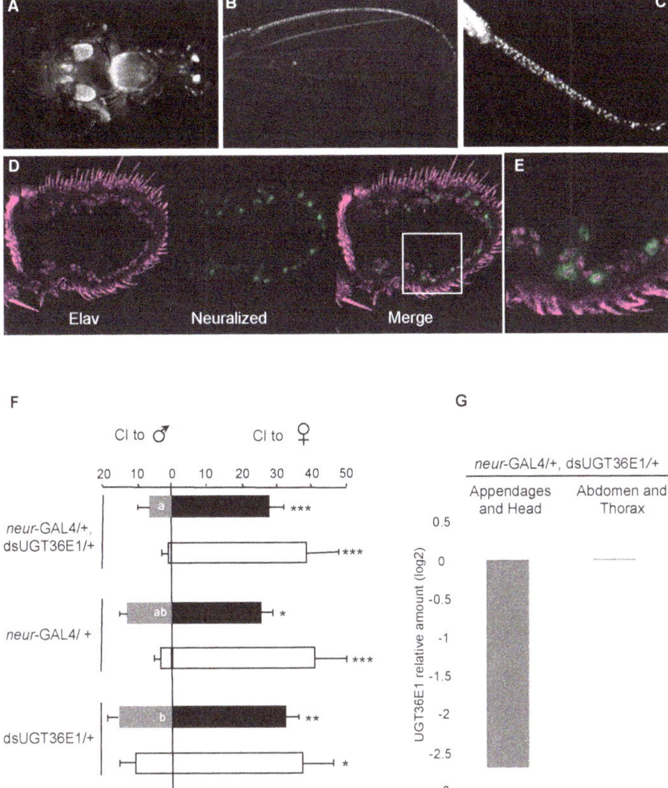

Figure 2. Effect of dsUGT36E1 in the peripheral chemosensory system. A strong expression of *neuralized*-GAL4 (*neur*-GAL4) was detected in (**A**) the antennae and the proboscis, (**B**) wing margins, (**C**) legs of adult flies. (**D**) Antenna stained for Elav protein (red) and anti-GFP (*neur*-GAL4). (**E**) Magnified view of antennal neurons expressing Elav and *neur*-GAL4. (**F**) Male ability to discriminate sex partners in "*neur*-GAL4/+, dsUGT36E1/+" testers and in both transgenic controls (*neur*-GAL4/+ and dsUGT36E1/+) under red light (filled bars) and white light (empty bars). For each test, $n > 35$. (**G**) Real-time PCR analysis showing UGT36E1 mRNA level in different tissues of "*neur*-GAL4/+, dsUGT36E1/+" flies. The significant difference in transcript level ratio is based on a comparison between control (dsUGT36E1/+) and "*neur*-GAL4/+, dsUGT36E1/+" genotypes (pair-wise fixed reallocation randomization test). For statistics and conditions, see Figure 1. *** $p < 0.001$; ** $p < 0.01$; * $p < 0.05$.

Next, we targeted subsets of peripheral chemosensory neurons potentially involved in pheromonal perception. Given that "*neur*-GAL4/+, dsUGT36E1/+" males showed reduced CIm, we targeted the dsUGT36E1 transgene in Gr66a gustatory sensory neurons which are involved in the detection of a male aversive pheromone [62,63]. The CIf/CIm performance of transgenic males was not different compared to controls (Figure S3), indicating that UGT36E1 expression in Gr66a-expressing neurons is not required for sex discrimination. Differently, when dsUGT36E1 was targeted in the majority of peripheral olfactory sensory neurons (OSNs) using the *Orco*-GAL4 driver, manipulated males ("*Orco*-GAL4/+, dsUGT36E1/+") showed a higher discrimination ability as compared to controls (Figure 3A). This effect was due both to (i) the increased CIf of manipulated males compared to *Orco*-Gal4/+ control ($p = 0.0003$) and to (ii) their decreased CIm compared to dsUGT36E1/+ males ($p = 0.0034$). Note that *Orco*-Gal4/+ control transgenic males, which showed a wild-type-like discrimination, had a significantly decreased sexual activity to target females and/or to target males compared to several other control transgenic males (Figure S4). While the potential alteration induced by GAL4 in some chemosensory tissues has already been reported [62], this finding supports the idea that male sexual activity and sex discrimination can be affected separately. The role of UGT36E1 in sex pheromones detection provides also the first functional evidence of the involvement of an ODE in chemosensory neurons. We cannot exclude the possibility that non-neuronal accessory cells which also express ODEs [64,65] can additionally modulate the male sex pheromone(s) perception. Moreover, our data only provides an indirect evidence of the UGT36E1 expression in head olfactory appendages since we obtained no signal using an antibody specifically designed against this protein.

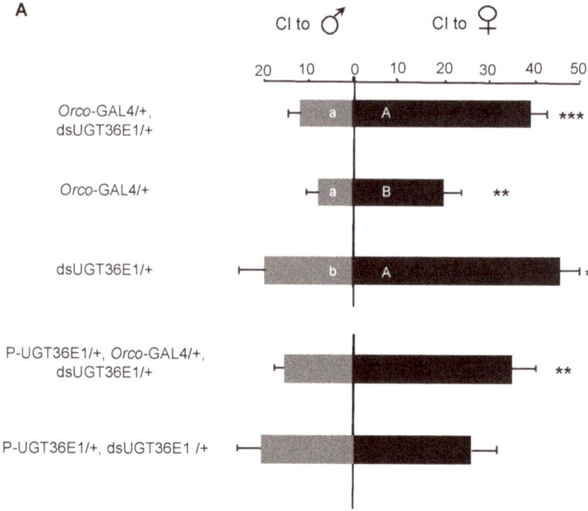

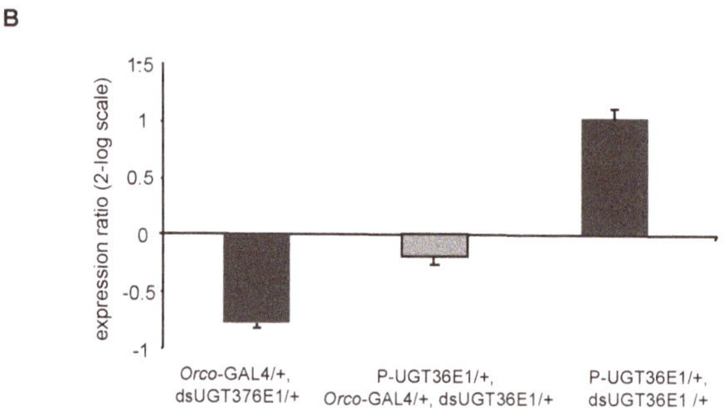

Figure 3. Expression and effect of dsUGT36E1 targeted in various sensory neurons subsets. (**A**) Targeting the dsUGT36E1 transgene in most olfactory sensory neurons with the *Orco*-GAL4 transgene ("*Orco*-GAL4/+, dsUGT36E1/+") improved mate choice performance compared to both transgenic controls. A similar targeting of this transgene in the P-UGT36E1 mutant background (in "*Orco*-GAL4/+, dsUGT36E1/+", but not "P- UGT36E1/+, dsUGT36E1/+") rescued male performance. For each behavioral test, $n > 35$. All courtship tests were carried out under red light. (**B**) Quantitative analysis of UGT36E1 expression in the sensory appendages of "P-UGT36E1/+, *Orco*-GAL4/+, dsUGT36E1/+" flies compared to "*Orco*-GAL4/+, dsUGT36E1/+" and "P-UGT36E1/+, dsUGT36E1/+" control flies. The wild-type strain Dijon was used as a reference. For statistics and conditions, see Figure 1. *** $p < 0.001$; ** $p < 0.01$; * $p < 0.05$.

Given the reciprocal effects induced by the P-UGT36E1 mutation and by the dsUGT36E1 RNAi transgene, both on UGT36E1 mRNA level and sexual discrimination, we combined the two genetic tools in the same fly. Strikingly, "P-UGT36E1/+, *Orco*-GAL4/+, dsUGT36E1/+" males showed a wild- type-like discrimination ability, whereas control "P-UGT36E1/+, dsUGT36E1/+" males showed no such preference (Figure 3A). This indicates that the dsUGT36E1 RNAi (increasing sex discrimination) compensated for the behavioral defect caused by the P-UGT36E1 mutation (decreasing sex discrimination). Moreover, UGT36E1 mRNA levels, measured in the fly appendages, did not differ between "P-UGT36E1/+, *Orco*-GAL4/+, dsUGT36E1/+" males, on one hand and wild-type and transgenic control flies on the other (Figure 3B). These experiments strongly suggest that the gene mutation and the RNAi transgene have additive effects both at the molecular and behavioral levels.

If we assume that both mRNA and UGT enzyme levels are correlated, our data suggest that the ability of male flies to discriminate sex pheromones depends on the UGT expression level in antennal OSNs. Compared to wild-type males, a reduction of UGT expression level in OSNs tends to increase male ability to discriminate sex pheromones, while a higher level induces the opposite effect. This suggests that the expression level of the UGT gene product in wild-type flies is somewhat intermediate between the levels in the P-UGT36E1 mutant and in "*Orco*-GAL4/+, dsUGT36E1/+" males. Such intermediate level may reflect a trade-off between a relatively high non-specific expression for optimal detoxification and/or signal termination and a relatively low expression allowing wild-type male to acutely detect and discriminate pheromonal stimuli.

To further investigate the function of UGT36E1 in the peripheral olfaction of sex pheromones, we recorded the global electrophysiological responses of individual male fly antennae (electroantennogram = EAG) either stimulated by male or female volatile pheromonal compounds. We used EAG instead of single sensilla recordings (SSRs) given that we had no idea of the sensillum or sensilla which could respond to the pheromonal mixture. Responses to sex-specific stimuli were normalized with 2-H, a general odorant eliciting robust antennal responses. The responses to these three olfactory stimuli were compared between wild-type, P-UGT36E1 mutant and "*Orco*-GAL4/+, dsUGT36E1/+" males (and in their transgenic parental controls). Wild-type male antenna showed slightly larger relative responses to male than to female volatile compounds ($p = 0.014$; Figure 4A,B). In P-UGT36E1 mutant males, the difference of relative responses to male and female volatile compounds strongly increased ($p = 0.0006$). This increased difference was due to a significantly increased response to male volatile compounds in the mutant compared to wild-type males ($n = 11–15$; $p = 0.02$). However, both mutant and wild-type males showed similar relative responses to female volatile compounds. On the other hand, "*Orco*-GAL4/+, dsUGT36E1/+" experimental males showed similar relative responses to female and male volatile compounds whereas parental transgenic controls (dsUGT36E1/+ and *Orco*-GAL4/+) showed a wild-type-like pattern (e.g., a slightly larger relative response to male than to female volatile compounds; Figure 4C,D). These results indicate that the P-UGT36E1 mutation enhanced the relative amplitude of the electrophysiological response to male volatile compounds whereas the RNAi directed against UGT36E1 in OSNs reduced this response. Therefore, if we cannot formally rule out the possibility that the UGT also affected the EAG response to 2-H, our data clearly show that the manipulation of the UGT gene affected the EAG response to male pheromone(s) relative to female pheromone(s).

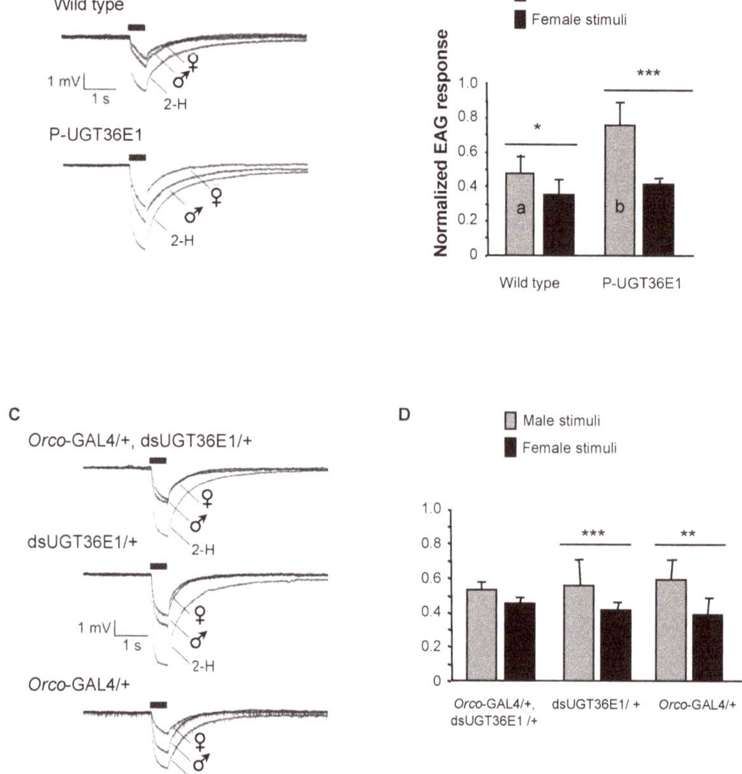

Figure 4. Electrophysiological recording of male olfactory response to different stimuli. (**A**) The graphs represent averaged electroantennogram (EAG) responses of wild-type and P-UGT36E1 mutant male flies to three stimuli: living females, living males and 2-Heptanone (2-H). The thick bars indicate the stimulus duration. (**B**) Bars represent the EAG responses to males and females normalized with the respective responses to 2-H. Statistical differences were noted (i) for the significance of the difference to both sex stimuli (above each pair of bars) and (ii) for responses to male stimuli (letters inside lightly filled bars), between genotypes. (**C**) Averaged EAG responses for "*Orco*-GAL4/+, dsUGT36E1/+", "dsUGT36E1/+" and "*Orco*-GAL4/+" males presented with the same three stimuli. (**D**) Normalized EAG responses to males and females. "*Orco*-GAL4/+, dsUGT36E1/+" males showed no difference in their responses to either sex. For each test, $n > 11$. For statistics and conditions, see Figure 1. *** $p < 0.001$; ** $p < 0.01$; * $p < 0.05$.

Involvement of ODEs in pheromonal signal modulation was previously hypothesized [20,35,36,66–68]. Here we propose that the abundance of the UGT36E1 enzyme in *Drosophila* OSNs can affect the clearance of pheromones in the perireceptor space. Based on this hypothesis, the over-expression of the gene (in P-UGT36E1 mutants) would increase the clearance of the stimulus, promoting faster successive stimulations, and thus enhance the effect of the male inhibitory pheromone (as measured by the EAG), this leading to reduced sex discrimination. Reciprocally, the decreased UGT expression in RNAi targeted males could affect the pheromonal clearance, resulting in an overstimulation which negatively impacts the signal level. This would reduce the aversive effect induced by the male pheromone(s) and increase male ability to discriminate sex pheromones. Based on these observations, we hypothesize that decreased pheromonal clearance in the perireceptor space promoted the saturation of dedicated receptors and reduced both OSN sensitivity and relative EAG amplitude. In support of our interpretation, two other studies based on the detection of different chemical or use of different genetic tools, reported that the alteration of activity in phase I enzymes (CYP or CES) also induced a prolonged olfactory neuronal response leading to an altered perception of the pheromones highlighting their role in signal termination [7,36].

How can we explain the apparent conundrum between the increased EAG response to male pheromone and the decreased discrimination shown by mutant males (and the reciprocal effects in RNAi targeted males)? In our behavioral assay, wild-type male OSNs were simultaneously stimulated by a mixture of inhibitory and attractive olfactory pheromones emitted by male and female flies, respectively. We hypothesize that the increased nervous antennal response to male inhibitory pheromone(s) in the perireceptor space may disturb the response to female pheromone(s), and this unbalanced effect would affect their integrated comparison in brain structures which are normally involved in the sex pheromone discrimination [69,70]. In any case, our data reveal that this mechanism likely depends on the level of the UGT gene product: males combining both the mutation and the RNAi targeted in neural tissues showed a wild-type-like behavioral discrimination. This is reminiscent of a recent transcriptomic study performed in the *Bombyx mori* silkworm antenna which revealed that the olfactory impairment observed in the domestic strain is correlated with a decreased expression of ODEs (including some UGTs) as compared to the wild *B. mori* strain [15].

In summary, our data reveal that the reciprocal variation of a UGT expression can change male sex discrimination in opposite directions. This effect is likely based on the ability of manipulated flies to discriminate between volatile sex pheromones. Given the high diversity, the ubiquitous distribution, the regulation and the varied properties of ODEs in animals, our findings provide a significant step to unravel the complexity of mechanisms underlying olfactory sensitivity at the peripheral nervous system.

Supplementary Materials: The following are available online at http://www.mdpi.com/2073-4425/11/3/237/s1, Figure S1: Expression ratio of UGT36E1 in appendages, head and thorax and abdomen, Figure S2: Locomotor activity tests, Figure S3: Effect of dsUGT36E1 targeted in Gr66a gustatory neurons, Figure S4: Male courtship in control strains.

Author Contributions: Conceptualization, Y.A., J.-F.F. and J.-M.H.; methodology, P.L., I.C., P.F., F.N., S.F. and A.L.; validation, J.-F.F. and J.-M.H.; formal analysis, S.F., A.L. and P.L.; investigation, S.F., A.L., P.L., P.F. and I.C.; writing—original draft preparation, S.F., A.L., Y.A., L.B., F.N., J.-M.H. and J.-F.F.; writing—review and editing, S.F., A.L., Y.A., L.B., F.N., J.-M.H. and J.-F.F.; supervision, J.-F.F. and J.-M.H.; project administration, J.-M.H.; funding acquisition, Y.A., L.B., J.-F.F. and J.-M.H. All authors have read and agreed to the published version of the manuscript.

Funding: This research was funded by Agence National de la Recherche (ANR-08-BLAN-0203) by the CNRS, INRA, the Burgundy Regional Council and the French Ministry of Higher Education, Research and Innovation.

Acknowledgments: We thank Matthew Cobb for comments on a previous version of the manuscript, Catherine Méart and Adrien François for their invaluable technical assistance and DIMACell plateform for microscopy studies (Université de Bourgogne Franche-Comté).

Conflicts of Interest: The authors declare no conflict of interest.

References

1. Getchell, T.V.; Margolis, F.L.; Getchell, M.L. Perireceptor and receptor events in vertebrate olfaction. *Prog. Neurobiol.* **1984**, *23*, 317–345. [CrossRef]
2. Vogt, R.G.; Riddiford, L.M. Pheromone binding and inactivation by moth antennae. *Nature* **1981**, *293*, 161–163. [CrossRef] [PubMed]
3. Scheuermann, E.A.; Smith, D.P. Odor-specific deactivation defects in a Drosophila odorant-binding protein mutant. *Genetics* **2019**, *213*, 897–909. [CrossRef] [PubMed]
4. Gonzalez, D.; Rihani, K.; Neiers, F.; Poirier, N.; Fraichard, S.; Gotthard, G.; Chertemps, T.; Maibeche, M.; Ferveur, J.F.; Briand, L. The Drosophila odorant-binding protein 28a is involved in the detection of the floral odour ss-ionone. *Cell. Mol. Life Sci.* **2019**. [CrossRef]

5. Heydel, J.M.; Coelho, A.; Thiebaud, N.; Legendre, A.; Le Bon, A.M.; Faure, P.; Neiers, F.; Artur, Y.; Golebiowski, J.; Briand, L. Odorant-binding proteins and xenobiotic metabolizing enzymes: Implications in olfactory perireceptor events. *Anat. Rec. (Hoboken)* **2013**, *296*, 1333–1345. [CrossRef] [PubMed]
6. Heydel, J.M.; Faure, P.; Neiers, F. Nasal odorant metabolism: Enzymes, activity and function in olfaction. *Drug Metab. Rev.* **2019**, *51*, 224–245. [CrossRef]
7. Maibeche-Coisne, M.; Nikonov, A.A.; Ishida, Y.; Jacquin-Joly, E.; Leal, W.S. Pheromone anosmia in a scarab beetle induced by in vivo inhibition of a pheromone-degrading enzyme. *Proc. Natl. Acad. Sci. USA* **2004**, *101*, 11459–11464. [CrossRef]
8. Heydel, J.M.; Menetrier, F.; Belloir, C.; Canon, F.; Faure, P.; Lirussi, F.; Chavanne, E.; Saliou, J.M.; Artur, Y.; Canivenc-Lavier, M.C.; et al. Characterization of rat glutathione transferases in olfactory epithelium and mucus. *PLoS ONE* **2019**, *14*, e0220259. [CrossRef]
9. Gonzalez, D.; Fraichard, S.; Grassein, P.; Delarue, P.; Senet, P.; Nicolai, A.; Chavanne, E.; Mucher, E.; Artur, Y.; Ferveur, J.F.; et al. Characterization of a Drosophila glutathione transferase involved in isothiocyanate detoxification. *Insect Biochem. Mol. Biol.* **2018**, *95*, 33–43. [CrossRef]
10. Thiebaud, N.; Sigoillot, M.; Chevalier, J.; Artur, Y.; Heydel, J.M.; Le Bon, A.M. Effects of typical inducers on olfactory xenobiotic-metabolizing enzyme, transporter, and transcription factor expression in rats. *Drug Metab. Dispos. Biol. Fate Chem.* **2010**, *38*, 1865–1875. [CrossRef]
11. Younus, F.; Chertemps, T.; Pearce, S.L.; Pandey, G.; Bozzolan, F.; Coppin, C.W.; Russell, R.J.; Maibeche-Coisne, M.; Oakeshott, J.G. Identification of candidate odorant degrading gene/enzyme systems in the antennal transcriptome of Drosophila melanogaster. *Insect Biochem. Mol. Biol.* **2014**, *53*, 30–43. [CrossRef] [PubMed]
12. Mohapatra, P.; Menuz, K. Molecular profiling of the drosophila antenna reveals conserved genes underlying olfaction in insects. *G3 (Bethesda)* **2019**, *9*, 3753–3771. [CrossRef] [PubMed]
13. Yang, Y.; Li, W.; Tao, J.; Zong, S. Antennal transcriptome analyses and olfactory protein identification in an important wood-boring moth pest, Streltzoviella insularis (Lepidoptera: Cossidae). *Sci. Rep.* **2019**, *9*, 17951. [CrossRef] [PubMed]
14. Walker, W.B., III; Roy, A.; Anderson, P.; Schlyter, F.; Hansson, B.S.; Larsson, M.C. Transcriptome analysis of gene families involved in chemosensory function in Spodoptera Littoralis (Lepidoptera: Noctuidae). *BMC Genom.* **2019**, *20*, 428. [CrossRef]
15. Qiu, C.Z.; Zhou, Q.Z.; Liu, T.T.; Fang, S.M.; Wang, Y.W.; Fang, X.; Huang, C.L.; Yu, Q.Y.; Chen, C.H.; Zhang, Z. Evidence of peripheral olfactory impairment in the domestic silkworms: Insight from the comparative transcriptome and population genetics. *BMC Genom.* **2018**, *19*, 788. [CrossRef]
16. Durand, N.; Pottier, M.A.; Siaussat, D.; Bozzolan, F.; Maibeche, M.; Chertemps, T. Glutathione-S-transferases in the olfactory organ of the noctuid moth spodoptera littoralis, diversity and conservation of chemosensory clades. *Front. Physiol.* **2018**, *9*, 1283. [CrossRef]
17. Yang, S.; Cao, D.; Wang, G.; Liu, Y. Identification of genes involved in chemoreception in plutella xyllostella by antennal transcriptome analysis. *Sci. Rep.* **2017**, *7*, 11941. [CrossRef]
18. Wu, Z.; Zhang, H.; Bin, S.; Chen, L.; Han, Q.; Lin, J. Antennal and abdominal transcriptomes reveal chemosensory genes in the Asian citrus psyllid, diaphorina citri. *PLoS ONE* **2016**, *11*, e0159372. [CrossRef]
19. Wu, Z.; Bin, S.; He, H.; Wang, Z.; Li, M.; Lin, J. Differential expression analysis of chemoreception genes in the striped flea beetle phyllotreta striolata using a transcriptomic approach. *PLoS ONE* **2016**, *11*, e0153067. [CrossRef]
20. He, P.; Zhang, Y.F.; Hong, D.Y.; Wang, J.; Wang, X.L.; Zuo, L.H.; Tang, X.F.; Xu, W.M.; He, M. A reference gene set for sex pheromone biosynthesis and degradation genes from the diamondback moth, Plutella xylostella, based on genome and transcriptome digital gene expression analyses. *BMC Genom.* **2017**, *18*, 219. [CrossRef]
21. Hu, P.; Wang, J.; Cui, M.; Tao, J.; Luo, Y. Antennal transcriptome analysis of the Asian longhorned beetle Anoplophora glabripennis. *Sci. Rep.* **2016**, *6*, 26652. [CrossRef] [PubMed]
22. Liu, S.; Gong, Z.J.; Rao, X.J.; Li, M.Y.; Li, S.G. Identification of putative carboxylesterase and glutathione S-transferase genes from the antennae of the chilo suppressalis (Lepidoptera: Pyralidae). *J. Insect Sci.* **2015**, *15*. [CrossRef]
23. Leitch, O.; Papanicolaou, A.; Lennard, C.; Kirkbride, K.P.; Anderson, A. Chemosensory genes identified in the antennal transcriptome of the blowfly Calliphora stygia. *BMC Genom.* **2015**, *16*, 255. [CrossRef] [PubMed]

24. Corcoran, J.A.; Jordan, M.D.; Thrimawithana, A.H.; Crowhurst, R.N.; Newcomb, R.D. The peripheral olfactory repertoire of the lightbrown apple moth, epiphyas postvittana. *PLoS ONE* **2015**, *10*, e0128596. [CrossRef]
25. Choo, Y.M.; Pelletier, J.; Atungulu, E.; Leal, W.S. Identification and characterization of an antennae-specific aldehyde oxidase from the navel orangeworm. *PLoS ONE* **2013**, *8*, e67794. [CrossRef] [PubMed]
26. Durand, N.; Chertemps, T.; Maibeche-Coisne, M. Antennal carboxylesterases in a moth, structural and functional diversity. *Commun. Integr. Biol.* **2012**, *5*, 284–286. [CrossRef]
27. Merlin, C.; Rosell, G.; Carot-Sans, G.; Francois, M.C.; Bozzolan, F.; Pelletier, J.; Jacquin-Joly, E.; Guerrero, A.; Maibeche-Coisne, M. Antennal esterase cDNAs from two pest moths, Spodoptera littoralis and Sesamia nonagrioides, potentially involved in odourant degradation. *Insect Mol. Biol.* **2007**, *16*, 73–81. [CrossRef]
28. Merlin, C.; Francois, M.C.; Bozzolan, F.; Pelletier, J.; Jacquin-Joly, E.; Maibeche-Coisne, M. A new aldehyde oxidase selectively expressed in chemosensory organs of insects. *Biochem. Biophys. Res. Commun.* **2005**, *332*, 4–10. [CrossRef]
29. Maibeche-Coisne, M.; Merlin, C.; Francois, M.C.; Queguiner, I.; Porcheron, P.; Jacquin-Joly, E. Putative odorant-degrading esterase cDNA from the moth Mamestra brassicae: Cloning and expression patterns in male and female antennae. *Chem. Senses* **2004**, *29*, 381–390. [CrossRef]
30. Ishida, Y.; Leal, W.S. Cloning of putative odorant-degrading enzyme and integumental esterase cDNAs from the wild silkmoth, Antheraea Polyphemus. *Insect Biochem. Mol. Biol.* **2002**, *32*, 1775–1780. [CrossRef]
31. Rybczynski, R.; Vogt, R.G.; Lerner, M.R. Antennal-specific pheromone-degrading aldehyde oxidases from the moths Antheraea Polyphemus and Bombyx mori. *J. Biol. Chem.* **1990**, *265*, 19712–19715. [PubMed]
32. Pottier, M.A.; Bozzolan, F.; Chertemps, T.; Jacquin-Joly, E.; Lalouette, L.; Siaussat, D.; Maibeche-Coisne, M. Cytochrome P450s and cytochrome P450 reductase in the olfactory organ of the cotton leafworm Spodoptera littoralis. *Insect Mol. Biol.* **2012**, *21*, 568–580. [CrossRef] [PubMed]
33. Durand, N.; Carot-Sans, G.; Chertemps, T.; Bozzolan, F.; Party, V.; Renou, M.; Debernard, S.; Rosell, G.; Maibeche-Coisne, M. Characterization of an Antennal carboxylesterase from the pest moth Spodoptera littoralis degrading a host plant odorant. *PLoS ONE* **2010**, *5*, e15026. [CrossRef] [PubMed]
34. Ahn, S.J.; Dermauw, W.; Wybouw, N.; Heckel, D.G.; Van Leeuwen, T. Bacterial origin of a diverse family of UDP-glycosyltransferase genes in the Tetranychus urticae genome. *Insect Biochem. Mol. Biol.* **2014**, *50*, 43–57. [CrossRef]
35. Durand, N.; Carot-Sans, G.; Bozzolan, F.; Rosell, G.; Siaussat, D.; Debernard, S.; Chertemps, T.; Maibeche-Coisne, M. Degradation of pheromone and plant volatile components by a same odorant-degrading enzyme in the cotton leafworm, Spodoptera littoralis. *PLoS ONE* **2011**, *6*, e29147. [CrossRef]
36. Chertemps, T.; Francois, A.; Durand, N.; Rosell, G.; Dekker, T.; Lucas, P.; Maibeche-Coisne, M. A carboxylesterase, Esterase-6, modulates sensory physiological and behavioral response dynamics to pheromone in Drosophila. *BMC Biol.* **2012**, *10*, 56. [CrossRef]
37. Zhou, Y.; Fu, W.B.; Si, F.L.; Yan, Z.T.; Zhang, Y.J.; He, Q.Y.; Chen, B. UDP-glycosyltransferase genes and their association and mutations associated with pyrethroid resistance in Anopheles sinensis (Diptera: Culicidae). *Malar. J.* **2019**, *18*, 62. [CrossRef]
38. Zhao, J.; Xu, L.; Sun, Y.; Song, P.; Han, Z. UDP-Glycosyltransferase genes in the striped rice stem borer, chilo suppressalis (walker), and their contribution to chlorantraniliprole Resistance. *Int. J. Mol. Sci.* **2019**, *20*, 1064. [CrossRef]
39. Tian, F.; Wang, Z.; Li, C.; Liu, J.; Zeng, X. UDP-Glycosyltransferases are involved in imidacloprid resistance in the Asian citrus psyllid, Diaphorina citri (Hemiptera: Lividae). *Pestic. Biochem. Physiol.* **2019**, *154*, 23–31. [CrossRef]
40. Chen, X.; Xia, J.; Shang, Q.; Song, D.; Gao, X. UDP-glucosyltransferases potentially contribute to imidacloprid resistance in Aphis gossypii glover based on transcriptomic and proteomic analyses. *Pestic. Biochem. Physiol.* **2019**, *159*, 98–106. [CrossRef]
41. Wang, M.Y.; Liu, X.Y.; Shi, L.; Liu, J.L.; Shen, G.M.; Zhang, P.; Lu, W.C.; He, L. Functional analysis of UGT201D3 associated with abamectin resistance in Tetranychus cinnabarinus (Boisduval). *Insect Sci.* **2018**. [CrossRef] [PubMed]
42. Li, X.; Shi, H.; Gao, X.; Liang, P. Characterization of UDP-glucuronosyltransferase genes and their possible roles in multi-insecticide resistance in Plutella xylostella (L.). *Pest Manag. Sci.* **2018**, *74*, 695–704. [CrossRef] [PubMed]

43. Krempl, C.; Sporer, T.; Reichelt, M.; Ahn, S.J.; Heidel-Fischer, H.; Vogel, H.; Heckel, D.G.; Joussen, N. Potential detoxification of gossypol by UDP-glycosyltransferases in the two Heliothine moth species Helicoverpa armigera and Heliothis virescens. *Insect Biochem. Mol. Biol.* **2016**, *71*, 49–57. [CrossRef] [PubMed]
44. Ahn, S.J.; Badenes-Perez, F.R.; Reichelt, M.; Svatos, A.; Schneider, B.; Gershenzon, J.; Heckel, D.G. Metabolic detoxification of capsaicin by UDP-glycosyltransferase in three Helicoverpa species. *Arch. Insect Biochem. Physiol.* **2011**, *78*, 104–118. [CrossRef]
45. Bock, K.W. The UDP-glycosyltransferase (UGT) superfamily expressed in humans, insects and plants: Animal-plant arms-race and co-evolution. *Biochem. Pharmacol.* **2016**, *99*, 11–17. [CrossRef]
46. Heydel, J.M.; Holsztynska, E.J.; Legendre, A.; Thiebaud, N.; Artur, Y.; Le Bon, A.M. UDP-glucuronosyltransferases (UGTs) in neuro-olfactory tissues: Expression, regulation, and function. *Drug Metab. Rev.* **2010**, *42*, 74–97. [CrossRef]
47. Ahn, S.J.; Vogel, H.; Heckel, D.G. Comparative analysis of the UDP-glycosyltransferase multigene family in insects. *Insect Biochem. Mol. Biol.* **2012**, *42*, 133–147. [CrossRef]
48. Luque, T.; O'Reilly, D.R. Functional and phylogenetic analyses of a putative Drosophila melanogaster UDP-glycosyltransferase gene. *Insect Biochem. Mol. Biol.* **2002**, *32*, 1597–1604. [CrossRef]
49. Wang, S.; Liu, Y.; Zhou, J.J.; Yi, J.K.; Pan, Y.; Wang, J.; Zhang, X.X.; Wang, J.X.; Yang, S.; Xi, J.H. Identification and tissue expression profiling of candidate UDP-glycosyltransferase genes expressed in Holotrichia parallela motschulsky antennae. *Bull. Entomol. Res.* **2018**, *108*, 807–816. [CrossRef]
50. Bozzolan, F.; Siaussat, D.; Maria, A.; Durand, N.; Pottier, M.A.; Chertemps, T.; Maibeche-Coisne, M. Antennal uridine diphosphate (UDP)-glycosyltransferases in a pest insect: Diversity and putative function in odorant and xenobiotics clearance. *Insect Mol. Biol.* **2014**, *23*, 539–549. [CrossRef]
51. Wang, Q.; Hasan, G.; Pikielny, C.W. Preferential expression of biotransformation enzymes in the olfactory organs of Drosophila melanogaster, the antennae. *J. Biol. Chem.* **1999**, *274*, 10309–10315. [CrossRef] [PubMed]
52. Dietzl, G.; Chen, D.; Schnorrer, F.; Su, K.C.; Barinova, Y.; Fellner, M.; Gasser, B.; Kinsey, K.; Oppel, S.; Scheiblauer, S.; et al. A genome-wide transgenic RNAi library for conditional gene inactivation in Drosophila. *Nature* **2007**, *448*, 151–156. [CrossRef] [PubMed]
53. Everaerts, C.; Cazalé-Debat, L.; Louis, A.; Pereira, E.; Farine, J.P.; Cobb, M.; Ferveur, J.F. Larval imprinting alters adult sex pheromone response in Drosophila. *PeerJ* **2018**, *6*, e5585. [CrossRef] [PubMed]
54. Greenspan, R. *Fly Pushing: The Theory and Practice of Drosophila Genetics*, 2nd ed.; Cold Spring Harbor Laboratory Press: Cold Spring Harbor, NY, USA, 2004; p. 191.
55. Marcillac, F.; Grosjean, Y.; Ferveur, J.F. A single mutation alters production and discrimination of Drosophila sex pheromones. *Proceedings* **2005**, *272*, 303–309. [CrossRef] [PubMed]
56. Ferveur, J.F.; Sureau, G. Simultaneous influence on male courtship of stimulatory and inhibitory pheromones produced by live sex-mosaic Drosophila melanogaster. *Proceedings* **1996**, *263*, 967–973.
57. Pfaffl, M.W. A new mathematical model for relative quantification in real-time RT-PCR. *Nucleic Acids Res.* **2001**, *29*, e45. [CrossRef] [PubMed]
58. Krishnan, P.; Dryer, S.E.; Hardin, P.E. Measuring circadian rhythms in olfaction using electroantennograms. *Methods Enzymol.* **2005**, *393*, 495–508.
59. van der Goes van Naters, W.; Carlson, J.R. Receptors and neurons for fly odors in Drosophila. *Curr. Biol.* **2007**, *17*, 606–612. [CrossRef]
60. Pfaffl, M.W.; Horgan, G.W.; Dempfle, L. Relative expression software tool (REST) for group-wise comparison and statistical analysis of relative expression results in real-time PCR. *Nucleic Acids Res.* **2002**, *30*, e36. [CrossRef]
61. Ferveur, J.F. Cuticular hydrocarbons: Their evolution and roles in Drosophila pheromonal communication. *Behav. Genet.* **2005**, *35*, 279–295. [CrossRef]
62. Lacaille, F.; Hiroi, M.; Twele, R.; Inoshita, T.; Umemoto, D.; Maniere, G.; Marion-Poll, F.; Ozaki, M.; Francke, W.; Cobb, M.; et al. An inhibitory sex pheromone tastes bitter for Drosophila males. *PLoS ONE* **2007**, *2*, e661. [CrossRef] [PubMed]
63. Moon, S.J.; Lee, Y.; Jiao, Y.; Montell, C. A Drosophila gustatory receptor essential for aversive taste and inhibiting male-to-male courtship. *Curr. Biol.* **2009**, *19*, 1623–1627. [CrossRef] [PubMed]
64. Younus, F.; Fraser, N.J.; Coppin, C.W.; Liu, J.W.; Correy, G.J.; Chertemps, T.; Pandey, G.; Maibeche, M.; Jackson, C.J.; Oakeshott, J.G. Molecular basis for the behavioral effects of the odorant degrading enzyme Esterase 6 in Drosophila. *Sci. Rep.* **2017**, *7*, 46188. [CrossRef] [PubMed]

65. Wang, L.; Dankert, H.; Perona, P.; Anderson, D.J. A common genetic target for environmental and heritable influences on aggressiveness in Drosophila. *Proc. Natl. Acad. Sci. USA* **2008**, *105*, 5657–5663. [CrossRef]
66. Feng, B.; Zheng, K.; Li, C.; Guo, Q.; Du, Y. A cytochrome P450 gene plays a role in the recognition of sex pheromones in the tobacco cutworm, Spodoptera litura. *Insect Mol. Biol.* **2017**, *26*, 369–382. [CrossRef]
67. He, P.; Li, Z.Q.; Liu, C.C.; Liu, S.J.; Dong, S.L. Two esterases from the genus Spodoptera degrade sex pheromones and plant volatiles. *Genome* **2014**, *57*, 201–208. [CrossRef]
68. Ishida, Y.; Leal, W.S. Rapid inactivation of a moth pheromone. *Proc. Natl. Acad. Sci. USA* **2005**, *102*, 14075–14079. [CrossRef]
69. Datta, S.R.; Vasconcelos, M.L.; Ruta, V.; Luo, S.; Wong, A.; Demir, E.; Flores, J.; Balonze, K.; Dickson, B.J.; Axel, R. The Drosophila pheromone cVA activates a sexually dimorphic neural circuit. *Nature* **2008**, *452*, 473–477. [CrossRef]
70. Ruta, V.; Datta, S.R.; Vasconcelos, M.L.; Freeland, J.; Looger, L.L.; Axel, R. A dimorphic pheromone circuit in Drosophila from sensory input to descending output. *Nature* **2010**, *468*, 686–690. [CrossRef]

Review

Comprehensive History of *CSP* Genes: Evolution, Phylogenetic Distribution and Functions

Guoxia Liu [1,†], Ning Xuan [1,†], Balaji Rajashekar [2], Philippe Arnaud [3], Bernard Offmann [3] and Jean-François Picimbon [1,4,*]

1 Biotechnology Research Center, Shandong Academy of Agricultural Sciences, Jinan 250100, China; girlgx@sina.com (G.L.); xuanning3205613@sina.com (N.X.)
2 Institute of Computer Science, University of Tartu, Tartu 50090, Estonia; balajior@gmail.com
3 Protein Engineering and Functionality Unit, University of Nantes, 44322 Nantes, France; philippe.arnaud@univ-nantes.fr (P.A.); bernard.offmann@univ-nantes.fr (B.O.)
4 School of Bioengineering, Qilu University of Technology, Jinan 250353, China
* Correspondence: jfpicimbon@163.com; Tel.: +86-531-89631190
† Same contribution: G.L. (Bemisia), Ning Xuan (Bombyx).

Received: 10 February 2020; Accepted: 6 April 2020; Published: 10 April 2020

Abstract: In this review we present the developmental, histological, evolutionary and functional properties of insect chemosensory proteins (CSPs) in insect species. CSPs are small globular proteins folded like a prism and notoriously known for their complex and arguably obscure function(s), particularly in pheromone olfaction. Here, we focus on direct functional consequences on protein function depending on duplication, expression and RNA editing. The result of our analysis is important for understanding the significance of RNA-editing on functionality of *CSP* genes, particularly in the brain tissue.

Keywords: tandem duplication; RNA mutation; adaptive process; lipid transport; xenobiotic resistance; neuroplasticity

---

This report reviews duplication, expression, evolution and RNA editing of *CSP* genes for neofunctionalization in insecticide resistance and neuroplasticity, with a particular special interest in functional properties of insect chemosensory proteins (CSPs).

Noticing that this gene family exhibits signs of RNA editing, we speculate that they play a role in interacting with diverse compounds, including mainly xenobiotics, lipids and fatty acids of the linoleic acid pathways. We do not attempt to give justice to the eluding nature of this protein family, but we attempt to address all the known aspects of the *CSPs* from genomic organization to expression analysis, which are perhaps important signature motifs of the multifunction. Accordingly, we report about gene duplication, ubiquitous expression in the whole insect body, expression in response to the application of insecticide, and new phylogenetic analyses before formulating a theory on the role of the pleiotropic nature of this protein gene family, which might be particularly important in pathways of cellular metabolism that regulate not only the immune system and digestive tract, but also the peripheral nervous system and brain.

In this study, we describe the genomic organization, chromosomal localization and gene structure of *CSPs* in the *Apis*/*Nasonia* model and a comparative analysis with *Bombyx*, *Pediculus* and *Tribolium* genomes, from which first genetic data about *CSPs* have been obtained. The choice to direct CSP research towards hymenoptera, in particular behaviors of bees and wasps, resides in the differences in the sensitivity of these insects to pesticides, social molecular pathways and the evolution of insect societies, as well as the complexity of adaptation and learning capacity. While many bee species are endangered on the brink of extinction as a result of excessive use of pesticides of all sorts, challenging pollination, biodiversity and environmental fate, the bee brain remains an exceptional model to see how insects can learn to associate odors and colors in a similar way to humans.

1. Introduction: The Family of Chemosensory Proteins (CSPs)

1.1. A Very Ancient Malleable Protein

Chemosensory proteins are a class of small (10–12 kDa) soluble proteins reported for the first time by Nomura et al. (1982) as an up-regulated factor in the regenerating legs of *Periplaneta americana* [1]. Soon enough, the same protein was identified in the antennae and legs from sexually mature adult cockroaches with some apparent differences between females and males, rather suggesting a "*chemodevol*" function for this protein, i.e., contributing both to tissue development and the recognition of sex-specific signals such as sex pheromones [2–6]. They fold into a flexible prism constituted of six alpha-helices, a hydrophobic inner side particularly suitable for the transport of long aliphatic chains and for specific conformational changes on ligand binding [7–11].

However, multifunction in CSPs, as coined by Picimbon (2003), mainly refers to the ubiquitous tissue-distribution of this protein family from non-sensory organs such as the gut and fat body in the internal abdominal area to the sensory structures of the antennae, palps and legs [12]. In the silkworm moth *Bombyx mori*, most CSPs are found to be co-expressed in the pheromone gland and to be up-regulated in several tissues following insecticide exposure [13,14]. Coincidentally, CSPs are found to be crucial transporters not only for molecules as diverse as fatty acid lipids such as linoleic acid, but also for insecticide xenobiotics of plant oil origin such as cinnamaldehyde, as recently described in the sweetpotato whitefly *Bemisia tabaci* [15]. Such patterns in tissue-distribution and such ligand diversity invite us to debate further about the complexity and/or multi-functional aspects of this more and more challenging protein family.

Complexity and multi-function in CSPs are largely brought by recent findings about DNA and RNA- polymerization, i.e., specific genetic events on the DNA/RNA template encoding CSP and structure variations [7–11,16,17]. The genetic code dictates the sequence of amino acids in CSP protein, but multiple variant isoforms can exist for a CSP thanks to various post-transcriptional events from intron splicing on RNA in the nucleus to editing of peptide molecules in the ribosome (Figure 1). For instance, using SWISS-MODEL, a software for homology modeling of protein structures and complexes [18], shows that a gene such as *BmorCSP4* (3 exons-2 introns) can lead to eleven protein subtypes, which can be used as templates to produce even more protein subtype variants following various genetic events such as RNA splicing, removal of intron, exon shuffling, RNA editing, protein recoding and protein re-arrangement (Figure 1). Gene, RNA and protein editing could account for the theoretical problem of CSPs interacting with a million or trillion possible ligands, referring not necessarily to the olfactory receptor combinatorial coding theory, but to the recognition, transport and degradation of an enormous amount of potential toxicants and/or all the lipid metabolites that are necessary to activate nuclear receptors, trigger enzymes in different reactions of a chain and/or regulate gene expression in various cellular physiological systems (Figure 1). All of these complementary mechanisms would enable the CSPs to be malleable, i.e., to have a sequence that can be recoded in order to orientate the protein to a new function. The CSP malleability might be crucial, not only for a metabolic tissue such as the gut or the fat body, which certainly needs to degrade a million different xenobiotic chemicals of all sorts, but also for a multi or toti-potent cell on a way to transform in a multitude of organs and tissues or for the multiple synapses establishing new connections in the central and peripheral nervous systems [19].

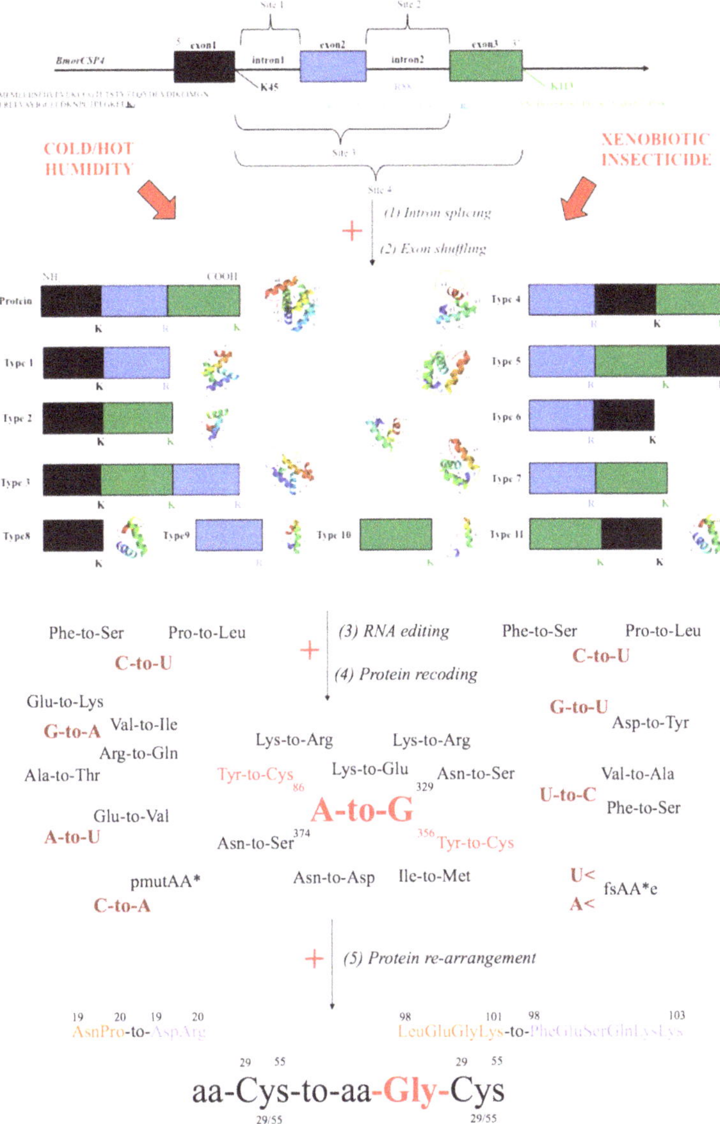

Figure 1. Gene splicing and RNA editing mechanism for diversification of chemosensory proteins (CSPs) under environmental change. An insect *CSP* gene structure such as lepidopteran *BmorCSP4* contains 2 introns and 3 exons [13,14]. (1) Genomic DNA (bold black line) is transcribed into premature mRNA yielding four possible sites for intron splicing. (2) The native protein sequence and eleven types of protein sequence variants can be produced by intron splicing, excision of non-coding regions (intron boundaries: K45, R88 and K113) and shuffling of coding regions (exon1 in black, exon2 in blue and exon 3 in green box). The folded shape of BmorCSP4 and a number of 11 new protein foldings (11 variants) can be generated from gene splicing. (3) The primary transcripts that are a faithful copy of the gene and variant mRNAs are all subject to further typo RNA editing, resulting in an increased number of genetic variants and protein subtypes [13,16–20]. Each mRNA is subject to mutations (A-to-G, A-to-U,

C-to-A, C-to-U, G-to-A, G-to-U and/or U-to-C) depending on external conditions (cold/hot temperature, humidity and/or exposure to xenobiotic insecticides). (4) The substitutions A-to-G at positions 86 and 356 build proteins harboring tyrosine (Tyr) to Cysteine (Cys) mutations in two different regions of BmorCSP4. Base deletion mutations (A< and U<) result in an early stop codon (fsAA*e), thereby yielding shortened proteins. C-to-A mutation changes the position of the stop codon (pmutAA*) and enhances the number of truncated protein isoforms/edited variants [13]. (5) The protein is recomposed not only after the translation process, i.e., when mRNA is translated to produce a protein, but also after protein synthesis. Once the protein is synthesized, the Asparagine-Proline (Asn-Pro) motif switches to another amino acid motif, Aspartate-Arginine (Asp-Arg). The Leucine-Glutamate-Glycine-Lysine (LeuGluGlyLys) motif changes to Phenylalanine-Glutamate-Serine-Glutamate-Lysine-Lysine (PheGluSerGluLysLys) in the C-terminal tail. A Glycine residue (Gly) is inserted next to Cysteine at position 29, 55 or both [13,14,16,17,19,20]. Protein structures are generated by BmorCSP4 templates in SWISS-MODEL using the X-ray crystal structure of MbraCSPA6 (1kx9.1.A) as a reference model [7,18].

Four bases, twenty residues, six types of conversion and only four editing enzymes may not be sufficient to underlie the extremely high number of protein variants described in *CSPs* as in the case of *Dscam*, ion channel and cochlear sensory genes [19,21–25]. Genetic variation via splicing and editing mechanisms in immune, neurobiological or sensory systems is probably needed to cause changes in protein families required for the recognition and transport of dozens, hundreds, thousands or millions of potential ligands [26,27]. In particular, we attempt to provide a comprehensive theoretical framework to explore the question of whether different edited versions of a protein such as CSP can be produced to cope with a wide variety of ligands, such as lipids and fatty acids, as well as drug compounds, insecticides and other xenobiotics. It is a hypothesis that is largely compatible with the existence of *CSPs* in different levels within many various kingdoms that contain organisms with cell walls, i.e., arthropods, bacteria, insects and plants [20,28–30].

1.2. CSPs and Cell Evolution

The *CSP* gene family is not specific to insects and other arthropod classes [20,28–30]. The existence of CSP in microbes cannot be a controversial issue. It is unlikely to see microbial samples contaminated by an arthropod tail, an insect scale or some leaf syrup. They are studied as strains that can cause serious infections in the lungs, blood and brain, so they are reared in very controlled areas in aseptic sterile clinical laboratories. It is therefore very unlikely to see an *Acinetobacter baumannii* RNA sample contaminated by a silkworm clone. Bacterial CSP (B-CSP)-RNA sequences have been found not only in Moraxellaceae *A. baumannii*, but also in Enterobacteriaceae *Escherichia coli*, Staphylococcaceae *Macrococcus caseolyticus*, Streptomycetaceae *Kitasatospora griseola*, *K. purpeofusca*, *K. sp. MBT66* and *K. sp. CB01950* and Xanthomonadaceae *Lysobacter capsici* (WP_043907137, WP_1212566, WP_071222707, WP_073810176/WP_083646628, WP_078880044, WP_082558797, WP_089438515, WP_096417339, WP_120787151, WP_120787152, WP_120787167 and WP_120787175) [29–31]. Very surprisingly, the very same proteins (BmorCSP2 and BmorCSP6) were found in *Bombyx* and in multi-species in bacteria. BmorCSPs were found not only in bacterial germs of the genus *Acinetobacter*, but also in *E. coli* [20,30,31]. This is an intriguing discovery to discuss function and evolution in the *CSP* gene family.

CSPs are highly conserved proteins, particularly in the Order Lepidoptera [6,12]. The presence of a same CSP sequence in some bacteria and in a few insect species such as the moths is enigmatic. It is very difficult to conceive that the very same sequence has been conserved for billions of years only in the clades bearing to a few insect genera. Such conserved proteins across two major divisions of life (bacteria and insects) may support the idea of a single universal common ancestor from which every life on earth or every new cell emerged. CSPs seem to be an intriguing coding part in the "dark matter" of the genomes of insects, worms, bacteria and yeasts. These genomes contain many highly conserved sequences whose functions are not yet known [32]. *CSPs* may be essential for the cell's organization, activity and adaptation, like many other gene families, including transfer RNAs and genes encoding the nucleotide-binding domain of ABC transporters [33]. If an identical protein sequence can be conserved along the evolution from bacteria to insects, it means that more CSP molecules or CSP-like proteins are to be found in the rest of the animal kingdom or that many various organisms have lost *CSP* at crucial steps during their evolutionary history. If the original protein encoded by *BmorCSP2* or *BmorCSP6* gene did not change by an iota despite horizontal gene transfer and the evolutionary change of cell and species over time, it might imply a key function in a basic common universal mechanism of eukaryote and prokaryote cells and in their interactions with an environment that continuously changes.

The presence of *CSPs* in plants so far appears to be a rather controversial point [34], particularly because plant samples can be easily contaminated by insect eggs, insect scales, insect feathers, many arthropods, fungi and/or bacteria. No efforts have been made to prove the existence of *CSP* in the plant genome. So far, only a rough analysis of plant EST database has been done, urging performing most accurate molecular biology work (molecular cloning of genomic DNA) in order to attest the occurrence of *CSPs* in plants as a fact [34]. This is essential to test the regulation of *CSP* expression in plant species of immense value as source of food or medicine under insecticide-contaminated soil. It could be that plants acquired *CSP* gene by horizontal transfer, but probably not from the insects. Most likely, horizontal gene transfer of *CSPs* occurred not only between microbes and insects or other arthropods, but also between diverse endosymbiotic microbial cells and a variety of plants [20,28–34].

1.3. Genome-Wide Identification, Comparative Genomics and Evolution of CSPs in Hymenoptera

The problem emerges that while insect CSP proteins are stated as being tuned to a high number of diverse functional ligands in many different insect physiological systems [7,8,15,35,36], the *CSP* gene family varies considerably in size across insect species such as moths (herbivorous lepidopteron) and body lice (hematophagous, phthirapteron). Moths retain twenty *CSPs*, while lice display only six *CSP* genes [14,31]. Similar to *Pediculus humanus corporis*, *Drosophila melanogaster* (carpophagous, dipteron) and *Anopheles gambiae* (hematophagous and nectariphagous, dipteron) have a rather low number of *CSPs* (the number of *CSPs (nb)* = 4–to–7 in these species) [12,31,37]. A significantly higher number of *CSP* genes are found in the genome of the red flour beetle, *Tribolium castaneum* (granivorous omnivorous, coleopteron) (*nb* = 19) [14,38]. This includes only model insect species where the genomic organization (clustering, grouping and mapping) of *CSPs* on the chromosomal level is known [12–14,31,37,38].

This poses the question of whether adapting and developing new phenotypes, the number of *CSPs* in insects and/or other arthropod species depends on the feeding habits and host preferences. Here, adding new sets of data in comparative genomics of a handful of species limited to two dipterons, one lepidopteron, one coleopteron and one phthirapteron were required to address hypothesis analysis. Therefore, we selected two cases in eusocial insect species (bees and wasps) to add two hymenopterons in the handful of species for which genome organization and structure of *CSPs* in insects are known [12–14,31,37,38]. We characterize hymenopteron's *CSPs*, annotation, classification, genomic organization, structure, phylogenetic distribution and expression of the *CSP* genes from honeybees (*Apis mellifera*) and parasitoid emerald jewel wasps (*Nasonia vitripennis*). In particular, we show that both *A. mellifera* and *N. vitripennis* have an extremely low number of *CSP* genes, as found, for instance, in the *Drosophila* fly, the *Anopheles* mosquito and the *Pediculus humanus corporis* louse (see Figure 2, Tables S1,S2) [12–14,31,37,38].

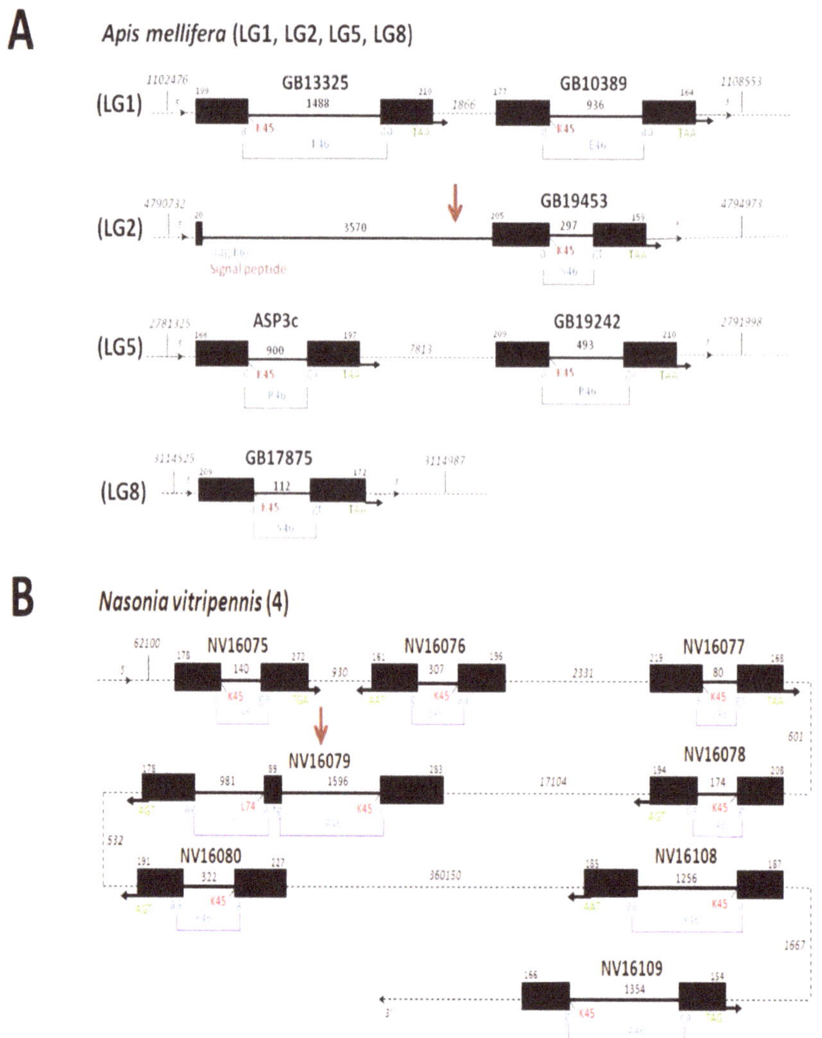

Figure 2. Genomic organization of *Apis* and *Nasonia CSPs*. (**A**) Genomic organization of *A. mellifera CSPs* on four different chromosomes (LG1, LG2, LG5 and LG8) (**B**) Genomic organization of *N. vitripennis CSPs* on chromosome 4. Exons are shown as black boxes, introns as bold black plain lines and intergenic intron regions as dotted lines. The numbers above the box and the plain line give the exon and intron size, respectively. The numbers in italics above the line give the distances between genes. Exon/intron sizes and intergenic distances are given in base pairs. The amino acid residue in red indicates the intron insertion site (K45, L74 or signal peptide). In blue is shown the triplet codon for the amino acid interrupted by intron insertion. Stop codons are indicated in green (*Apis*: TAA; *Nasonia*: TAA or TGA, A>G switch in stop codon). Horizontal arrow in black indicates the orientation of the gene: 5′–3′ (right) or 3′–5′ (left). The red vertical arrow points out the different position of the *double-intron CSP* gene in *A. mellifera* and *N. vitripennis*, respectively.

Although both species used as models, *A. mellifera* and *N. vitripennis*, do belong to the order Hymenoptera suborder Apocrita, they are part of two different clades, i.e., Aculeata and Parasitica that diverged more than 200 Mya. Key differences are found in their physical and behavioral characteristics. Parasitoid wasps such as *N. vitripennis* parasitize Diptera, mainly on the families Calliphoridae and Sarcophagidae. They seek out prey, kill the pupae they attack and lay eggs in the target host fly. So, the search of parasitic wasps is for oviposition. This behavior makes the wasps very distant from the honeybees that seek nectar and pollen from flowers and flowering plants. It is therefore of particular interest to compare their genomic variations in regards to *CSPs* to check whether *CSP* could be a mechanism by which behavioral or physiological characteristics of *Apis*, *Nasonia* and other model insect species have changed.

In the initial analysis of the honeybee genome, the existence of six *CSPs* has been reported in *A. mellifera*: six *single-intron* structures [37,39,40]. Analyzing a new assembly (sequence update) of the honeybee genome [41], we confirm the existence of only six *CSPs* in bees (*AmelASP3c*, *AmelGB10389*, *AmelGB13325*, *AmelGB17875*, *AmelGB19242* and *AmelGB19453*) and localize them on specific chromosomes (Figure 2A, Table S1). However, we find that there is an additional intron inserted in the signal peptide of *GB19453*. Bee *CSPs* are five *single-intron* structures and one *3 exons-two introns* structure (Figure 2A), not six *single-intron* structures as reported in the initial analysis [37,39,40]. In contrast to *A. mellifera CSPs*, no genomic data have ever been reported about *N. vitripennis CSPs*. Ten sequences encoding CSPs have been reported from the analysis of a cDNA library from the jewel wasp [42]. Here, we show that the number of ESTs encoding CSPs do not reflect the number of *CSPs*, as expected for a protein gene family with RNA variance like *CSP* genes. Here, we have performed a cautious genome analysis to precisely assess the number of genes encoding CSPs in the parasitoid jewel wasp [43]. Analyzing the assembly (sequence update) of the parasitoid wasp genome [43], we find only eight genes encoding CSPs: *NV46080*, *NV16108*, *NV16109*, *NV16075*, *NV16076*, *NV16077*, *NV16078* and *NV16079* (Figure 2B, Table S2). Therefore, after a cautious comparative analysis of the new assembly of *Apis* and *Nasonia* genomes, we report that these two hymenopteran species retain only six and eight *CSP* genes, respectively.

Interestingly, in contrast to their counterparts in *Tribolium* and *Bombyx*, *Apis* and *Nasonia CSPs* are all functional genes. In *Apis* and *Nasonia*, there are no pseudos or truncated *CSP* genes that have lost function after exon deletion [14,38] (Figure 2, Figure S1 and Tables S1,S2). Moreover, in contrast to coleopteran, dipteran and lepidopteran species, no intronless *CSP* genes are found in the honeybee *A. mellifera* and the parasitoid jewel wasp *N. vitripennis*, similar to the human body louse *P. humanus* [12–14,30,37,38] (Figure 2 and Tables S1,S2). Therefore, the data presented here argue for *CSP* loss as an essential evolutionary mechanism for adaptation and phenotypic variance not only in lice, but also in bees and parasitoid wasps.

Comparing *CSP* gene structures between *Apis* and *Nasonia*, we find that the number of *3 exons-2 introns (3e2i)* genes is the same (= 1), but the number of *single intron* genes is superior in *Nasonia* (+2; Figure 2 and Figure S1 and Tables S1 and S2). Silk moths have three *3e2i* genes among twenty *CSPs* [14]. The repertoire of *CSPs* in fruit flies is limited to two intronless and two small single intron genes. No *3e2i* gene structures are found among the four *CSPs from D. melanogaster* [12,37]. From the analysis published in Wanner et al., there are apparently also no *3e2i CSP* genes in the mosquito *A. gambiae* [37]. So, evidence in genomic analysis suggests that the number of *3e2i CSP* genes varies across insect species. Interestingly, beetles and lice are also known to retain only one *3e2i CSP* gene [30,38]. However, while beetles accumulated *CSPs* [38], the same arrangement of *CSPs* is maintained in lice and bees. *Pediculus CSPs* are one *3e2i* gene and five small *single intron* genes [30], as found for the honeybee (Figure 2A). Importantly, *Apis* strongly differs from *Nasonia* that shows a completely different genomic organization in *CSP* genes (see Figure 2). In *Apis*, they are clearly divided into four groups that are located on four different chromosomes (Figure 2A), while the *CSPs* are organized in the same small cluster of genes on chromosome 4 in *N. vitripennis* (Figure 2B). *GB13325/GB10389*, *GB19453*, *ASP3c/GB19242* and *GB17875* are located on chromosome LG1, LG2, LG5 and LG8, respectively (Figure 2A and Table S1). *GB13325* and *GB10389* are found near each other on LG1, while *ASP3c* and *GB19242* are found near each other on LG5. Furthermore, in bees, all *CSP* genes or pairs of *CSP* genes are found with the same transcriptional direction (5′–3′). They are about the same size (about 1 Kb) and differ only in intron size (from 112 to 3570 bps). Intron is always located at the same position, after Lys45, after the first base of the codon for amino acid at position 46 (Figure 2A). They all have TAA stop codon. Therefore, they might represent successive genome duplications, as described for the red flour beetle *T. castaneum* genome (beetlebase) [38] (see Figure 2A and Figure S1). Apparently, the duplicated copies of *GB19453* and *GB17875* were lost following genome duplication [41]; they are found as single genes on LG2 and LG8, respectively (Figure 2A).

On the contrary, *CSPs* occur in pairs and the members of each *CSP* pair are TGA and TAA-stop codon in the parasitoid jewel wasp *N. vitripennis*. Paired *CSP* genes are found in the opposite direction. The genes in the second group of *Nasonia CSPs* are oriented in an opposite direction (3′–5′) compared to *NV16079* (Figure 2B). Therefore, the *CSPs* from jewel wasps might originate from inverted gene duplication, which is in strong contrast with the *CSPs* from the bees.

This may be correlated with the position of *double intron* genes within the *CSP* gene cluster in *Apis* and *Nasonia*, respectively. In *Apis*, *GB19453* is located distantly from the other *CSPs* on a separated chromosome, while *NV16079* is located right in the middle of the *CSP* gene cluster on the same chromosome (chromosome 4) in *Nasonia* (Figure 2, red arrow). In the body of louse *P. humanus corporis*, the *double intron CSP* gene (*Phum594410*) is located farther away from the other *CSP* genes, as also found in *T. castaneum* [31,38]. This can also be found in *B. mori* where the three *3e2i CSPs* are located very distantly from each other [14]. Therefore, it is very unlikely that *3e2i CSP* genes come from the same common duplication event after analyzing the current handful of species (*Anopheles*, *Bombyx*, *Drosophila*, *Pediculus* and *Tribolium*) compared to our new data in *Apis* and *Nasonia*.

Furthermore, *Bombyx CSPs* are either TAG or TAA stop codon [13,14]. *Tribolium CSPs* are all TAA-stop codon [38]. *Pediculus CSP* genes are either TGA or TAA-stop codon, but not in pairs [31]. All these differences among stop codons, gene structures and genomic/chromosomal distributions show that *CSPs* from flour beetles, flies, moths, mosquitoes, lice and Hymenopteran species such as honeybees and jewel wasps have been subjected to different evolutionary paths that led to very specific genetic repertoires. This may reflect a unique evolutionary history for each insect lineage and suggest how the biology, the shape and the behavior exert strong influences on the evolution of the *CSP* repertoire.

Intron insertions occurred after the first base of the codon for amino acid 46, except for insertion in signal peptide (Figure 2). Interestingly, we find that in *CSP* genes such as *GB19453*, one intron is inserted only a few nucleotides after the start codon encoding the amino acid methionine (Figure 2A). The same observation (intron1 inserted shortly after the start of the signal peptide) was made in *AAJJ1196A* and *BmorCSP19* (Figure S1) [14,38,41–43]. In the case of these genes, the intron is inserted after the third base and therefore does not cause codon disruption (phase 0 intron). Phase 0 intron1 position suggests that splicing of the signal peptide region is tightly regulated and that the length of the signal peptide is functionally important in *CSPs*.

In addition, the intron is always inserted squarely in the middle of the *CSP* gene, between the two nucleotides that make up codon positions 1 and 2 in a specific codon that codes for amino acid 46 [14,30,37,38]. This is also observed in *CSPs* from honeybees and parasitoid wasps (Figure 2). In both species, the intron from *CSP* is located after the first base and disrupts the codon (phase 1 intron). Amino acid 46 can be Glu, Lys, Arg and Ser in the honeybee (Figure 2A). It can be Arg, Glu, Ser, Asp, Lys and Ala in the parasitoid wasp (Figure 2B). Therefore, it seems to be a widespread general view that the intron in a *CSP* gene contributes to the variability in amino acid 46 and requires very specific splicing mechanisms to avoid cutting a functional domain in CSP protein. Apparently, the intron boundaries of *CSPs* in many insect species indicate that the codon for amino acid 46 is a crucial site to underlie evolution and protein diversity in the *CSP* family.

Furthermore, we find that the insect genome seems to provide a simple form of sequence recovery (or backup). We find that the amino acid 46 is also coded by the three nucleotides at the tip of intron1 and intron2. All nucleotide combinations that code for amino acid 46 are found at the intron insertion site. Thus, *CSP* genes could be spliced at different codon positions without altering the primary amino acid composition of the CSP protein in any way. This is the case of *NV16077* where Ser46 can be encoded not only by *AGC* (disrupted codon), but also by *AGT* found at the intron1 boundary. This is also the case for *NV16079* where Arg75 can be encoded not only by *AGA* (disrupted codon), but also by *AGG* found in intron2 boundary. Therefore, a very important role is played by the codons at the intron boundary of *CSPs* to allow protein diversity.

Curiously, in our analysis, we find that there are no large introns containing a copy of gene or a retroposon in bee and wasp *CSPs*, in contrast to beetles and moths [14,38] (Figure 2, Table S1). The same situation has been described in human body lice [31]. *Pediculus CSP* genes (*PhumCSPs*) are all characterized by very short intron lengths (<288 bps) and all lack retroposon [31]. We find that the honeybee *A. mellifera* and the jewel wasp *N. vitripennis CSP* genes have introns varying in size between 80 and 3570 bps. The largest intron is intron2 from *GB19453* (Figure 2, Table S1). Importantly, *GB19453* and *NV16079* genes differ much, not only in intron size, but also in the position of intron boundaries. Introns in *GB19453* inserted after signal peptide and Lysine at position 45 (K45), respectively, while in *NV16079* they inserted after K45 and Arginine at position 75, respectively (Figure 2). This shows that despite a common exon–intron structure, these two genes do not originate from the duplication of a common ancestor, but rather from intron insertions that occurred independently in Apidae (honeybees) and Pteromalidae (parasitoids) during the course of evolution in the order Hymenoptera. Intron insertion also occurred independently in Lepidoptera as *BmorCSP10*, *BmorCSP14* and *BmorCSP19* show distinct intron boundaries (Figure S1) [14]. However, some specific *CSP* genes such as *AmelGB19453*, *AAJJ1796A*, *BmorCSP19* and *Phum594410* show the very same intron boundaries (intron1 inserted in signal peptide and intron 2 inserted after Lys45), strongly suggesting that insertion of intron1 in the signal peptide of *CSPs* occurred before the split of Hymenoptera, Coleoptera, Lepidoptera and Phthiraptera (parasites) [14,31,38] (see Figure 2A). There are no such *double-intron CSPs* in a parasitoid chalcid insect species such as the emerald jewel wasp *N. vitripennis* (see Figure 2B), which may indicate that this ancestral *double-intron CSP* gene was present in the last common ancestor of bees, beetles, moths and lice (i.e., more than 400 Mya), but was lost later during evolution in particular in parasitoids and other groups of predatory insects.

We reveal a high level of genetic plasticity in *CSPs*, which would be essential for evolutionary adaptation. This gene family is characterized by introns of different phases that inserted at different periods during the course of evolution in the insects. Some introns inserted at an early stage of evolution and were conserved even after the separation of the different insect lineages. Second intron inserted at a later stage of evolution, but was lost in some specific lineages, including the parasitoid lineage. We also reveal that gene duplication profiling within the *CSP* group is very different between honeybees (characterized by chromosomal duplication) and parasitoid wasps (characterized by extensive local inverted duplication), suggesting that the evolution of *CSP* genes may contribute to the development of very specific phenotypes and/or behavioral traits not only in hymenoptera, but also across many various organisms from bacteria to hymenoptera.

2. Phylogenetic Distribution Analysis in Insects and Bacteria

To measure the proportion of phenotypic variance attributable to genetic variance in *CSPs*, we analyzed the timeline of the evolutionary history of life from bacteria to insects and performed a phylogenetic analysis of the amino acid sequences using bacterial and insect *CSPs*. Our analysis shows that multiple duplications have taken place throughout the history of the gene family and eventually that, some of these duplications are unique to all hymenopteran species such as ants, bees and parasitoid wasps [44,45], while others are more ancient and are shared between various insect and bacterial orders (Figure 3).

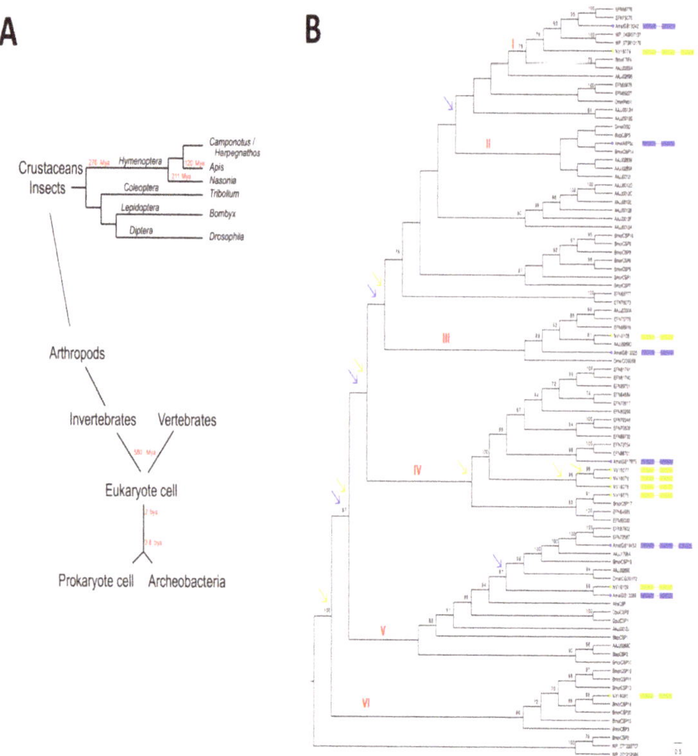

Figure 3. Schema of relationships from bacteria to insects and amino acid phylogenetic analysis of *Apis* and *Nasonia CSPs*. (**A**) Timeline of the evolutionary history of life from bacteria and prokaryote cells to multiple species of insects. (**B**) Gene phylogeny and orthology groups of bacterial/insect *CSPs* with focus on gene duplication profiling in honeybees and jewel wasps. Bacteria: *Acinetobacter* (WP_071212566, WP_071222707); *Kitasatospora* (WP_04307137, WP_07383810176) [29,30]. Insects: ants (EFN), beetles (AAJJ), flies (Dmel), moths (Bmor) and whiteflies (Btab) [12–15,37,38,44–46]. Crustacean: *A. franciscana* (AfraCSP; ABY62736, ABY62738); *D. pulex* (DpulCSP1, DpulCSP2; ABH88167, ABH88166). Phylogenetic trees are generated from a total of ninety protein sequences (IQ-TREE, UFBoot; 1000 replicates). Blue and green color circles represent *Apis mellifera* (*Amel*) and *Nasonia vitripennis* (*NV*) protein sequences, respectively. The gene structures are shown on the right for *Amel* (in blue) and *NV* (in green) *CSPs*. Branches are shown supported by >50% bootstrap value. Six major orthology groups are found corresponding to specific *Amel* and *NV* CSP sequences: group I (AmelGB19242, NV16079); group II (AmelASP3c); group III (AmelGB13325, NV16108); group IV (AmelGB17875, NV16075, NV16076, NV16077, NV16078); group V (AmelGB10389, AmelGB10453, NV16109); group VI (NV16080). Blue and green arrows indicate gene duplication profiling in *Amel* and *NV*, respectively. Supplementary Methods:Figure 3 The multiple sequence alignment was performed using Muscle global alignment (www.ebi.ac.uk/Tools/msa/muscle). Phylogenetic trees were constructed using IQ-TREE (http://iqtree.cibiv.univie.ac.at). The following parameters were used for phylogenetic tree construction, ultrafast bootstrap (UFBoot, using the –bb option of 1000 replicates), and a standard substitution model (-m TEST) was given for tree inference. The generated trees from IQ-TREE tool were visualized using Figtree (http://tree.bio.ed.ac.uk/software/figtree) and the branch-support values were recorded from the output treefile. The re-rooting was performed on WP_071212566 and WP_071222707 node. The trees were modified as cladogram and increasing order nodes were applied under trees section for better visualization.

In our phylogenetic analysis of *CSPs* from bacteria to insects, we also used the *CSPs* from *D. melanogaster* and *B. tabaci* as taxa since it was shown that dipteran and homopteran CSPs play a key role in insect defense [15,46]. Whiteflies such as *B. tabaci* show little in common with the pupal development of holometabolous insects (ants, bees, beetles, flies, moths and wasps). *Bemisia* is characterized by incomplete metamorphosis (hemimetabolous insect). The nymph resembles the adult in form and eating habits; there is no pupal stage in *B. tabaci*. The relationship of the bacterial and insect CSPs was studied with maximum parsimony (MP) analysis; MP was used to establish strict consensus trees using the IQ-TREE algorithm as described in Xuan et al. [16] (Figure 3).

In agreement with the phylogenetic distances between *Camponotus/Harpegnatos*, *Apis* and *Nasonia* (Figure 3A), our phylogenetic analysis shows that Hymenopteran CSPs such as EFN68779, EFN75075, AmelGB19242 and NV16079 are closely related; they form a group (group I) with a significant bootstrap value (78%; Figure 3B). Group I also includes *Bombyx* CSP4, *Tribolium* AAJJ0269A and two CSPs from the Streptomyces *Kitasatospora* (Figure 3B). The two CSP sequences from *Kitasatospora* bacterial strains (WP_04307137; WP_07383810176) fall close to NV16079 and AmelGB19242, showing a group of CSPs conserved from bacteria to insects. This group is clearly indicative of common ancestry between insect and bacterial *CSPs* [30,31]. However, AmelASP3c is more distantly related to this group I. AmelASP3c helps build another group of CSPs (group II), which also includes DmelOSD, BtabCSP3 (known to bind plant oil), BmorCSP14 and three Coleopteran CSPs, namely AAJJ0283B, AAJJ0283A and AAJJ0012I (Figure 3B).

The position on the tree (and gene structure) of *AmelGB19242* and *AmelASP3c* suggests that these two genes come from the same gene duplication that has happened before the split of Hymenoptera, Lepidoptera and Coleoptera, i.e., more than 300 Mya. Interestingly, there are no *Camponotus*, *Harpegnatos* or *Nasonia* clades in group II (Figure 3B), suggesting that this duplication event happened before the divergence of hymenopteran species, or that *AmelASP3c* gene has been lost in hymenopteran species such as the wood carpenter ants (*C. floridanus*), the predator jumping ants (*H. saltator*) or the parasitoid jewel wasps (*N. vitripennis*).

The wasp gene *NV16108* is clearly orthologous to Coleopteran *AAJJ0269D* gene (61% bootstrap). The two genes fall in a third group (group III) together with *AAJJ0330A*, *EFN75779*, *EFN66918*, *DmelCG9358* and *AmelGB13325* (89% bootstrap; Figure 3B). So, *NV16108* and *AmelGB13325* might originate from the same old gene duplication that took place in the far common ancestor of honeybees (Aculeata) and parasitoid wasps (Parasitica). In contrast, *NV16075*, *NV16076*, *NV16077* and *NV16078* might be the result of a series of much more recent gene duplications that specifically happened in Parasitica (group IV). These four genes seem to have been essential for the birth and evolution of the tiny parasitoid wasp, *N. vitripennis*. They labelled split-specific branches in *NV* with significantly high bootstrap values (96–98%; see green arrows, Figure 3B).

A larger group of *CSP* orthologs (group V) groups the honeybee genes *AmelGB19453* and *AmelGB10389* together with wasp *NV16109*, ant *EFN87902/EFN72587*, *Tribolium AAJJ1796A*, *AAJJ0269A* and *AAJJ0269E*, as well as the amino acid sequences for *Bombyx BmorCSP19* and *Drosophila DmelCG30172*. This group does not only include *CSP* genes from holometabolous insects, but includes also some genes expressed during the embryonic development in crustaceans (*AfraCSP*, *DpulCSP1* and *DpulCSP2*) and *B. tabaci chemosensory protein type 1* (*BtabCSP1*). BtabCSP1 is known to transport lipids such as linoleic acid (LA or C18:2 fatty acid) [15], suggesting that the main function of these CSPs from group V is to transport long fatty acid lipid chains such as C18:2. Using MP analysis, high bootstrap values (close to 100%) mean uniform support with BtabCSP1. All the characters informative enough to define group V agree that BtabCSP1 and other CSPs in this group are related with a common biological function (Figure 3B).

While *NV16109* and *AmelGB10389* are clearly two orthologous copies of the same gene (89% bootstrap value), *AmelGB19453* has no orthologous copy in *N. vitripennis*, begging the question of whether this absence is due to recent gene loss in some specific clades of the order Hymenoptera, similar to *AmelASP3c* (Figure 3B). *NV16080* forms an orthology group including several *BmorCSPs* but neither bee *CSP* family genes nor beetle *CSPs* are found in this group (group VI; Figure 3B). *BtabCSP1* and *BtabCSP2* (related to cinnamaldehyde transport) arose from a gene duplication that occurred in whiteflies, beetles and moths, but not in hymenopteran species [15,28] (Figure 3B). Therefore, the phenotype associated with plant-feeding habits and resistance to plant toxins seems to be associated with genetic variation, genetic changes, gene rearrangement and/or plasticity in some very specific groups of *CSPs*.

This poses the question of whether *CSPs* have contributed to the development of the eukaryote cell, the insect cell, as well as the bacterial prokaryote cell. The eukaryote cell divided into invertebrates and vertebrates about 580 Mya. The eukaryote cell was built about 2 Bya(Billion years ago), and the original archeobacterium and/or the prokaryote cell evolved 3.8 Bya (Figure 3A). The six orthology groups revealed in our phylogenetic analysis of bacterial/insect CSPs always display counterparts from various insect orders such as Coleoptera, Lepidoptera and/or Diptera, and in some cases, they also display a number of clades from the bacteria superkingdom. This indicates that *CSPs* originate from an extremely ancient duplication, which probably occurred prior to the origin of insects, much before the different insect orders took place (e.g., about >350–412 Mya). The first *CSP* gene duplication probably took place in some archeobacteria some billion years ago, perhaps approximately when life and diversity had to come from the original cell. So, the evolution and editing process in *CSPs* could date back to Bya and may eventually help develop an understanding of, not only cell fate and/or organismal evolution in various prokaryote systems, but also neural development and/or birth and evolution in highly diverse groups of eukaryote animal species. Interestingly, while it contains *AmelGB17875*, *BmorCSP17* and multiple copies of *Camponotus*, *Harpegnatos* and *Nasonia CSPs*, group IV lacks *Drosophila* and *Tribolium* clades, strongly suggesting that this gene has been subjected to continuous series of duplications in ants and parasitoid wasps, but has been lost specifically in Diptera and Coleoptera. Some *CSP* genes are more recent than others; some represent duplicates that occur specifically in Apocrita, but none of the six orthology groups that we describe here happen to be specific to bees, wasps or Hymenoptera (Groups I-VI; Figure 3B).

In addition, in our study, we find that the *3e2i/double-introns CSP* genes from *A. mellifera* and *N. vitripennis* (*AmelGB19453* and *NV16079*) group separately, confirming our first assumption that duplications as well as intron insertions have occurred independently in Aculeata (*Amel*) and Parasitica (*NV*), respectively (Figure 2; Figure 3). However, our most intriguing finding might be that bacterial *CSP* sequences such as *WP_071212566* and *WP_071222707* from Acinetobacter *A. baumannii* fall at the bottom of the phylogenetic tree, together with *NV16080* and multiple *CSP* sequences from the silkworm *B. mori* (*BmorCSP3*, *BmorCSP11*, *BmorCSP12*, *BmorCSP13*, *BmorCSP15*, *BmorCSP18* and *BmorCSP20*; Group VI: 86% bootstrap value). Importantly, we note that *B. mori* CSP2 sequence is identical to bacterial "CSPs" WP_071212566 and WP_071222707 (100% bootstrap), strongly suggesting that this set of proteins represents the most ancient form in the CSP family and an extremely old molecule, as well as being perhaps the most ancient type of carrier molecule in the earliest known life forms on Earth (back to >3 Bya; Figure 3AB). *CSP* gene duplicates can evolve so as to parse the original function or to acquire new roles. In our phylogenetic analysis, *NV16080* is orthologous to *BmorCSP18*, which is a truncated gene in the silkworm moth *B. mori* [14] (Figure 3B). So, it seems that the original function of this gene has been lost in silkworm, but multiple derived duplicated versions of *CSP18* have taken on the role of specifying moth identity. In contrast, loss of *NV16080* rather seems to have been decisive for the development of many other insect lineages such as ants, bees, beetles and flies (Figure 3).

Therefore, in our study, comparative genomics and phylogenetic analysis both show that *CSP* genes evolved through duplication and that many duplicated *CSP* genes had different fates as found not only in beetles, lice and moths, but also in honeybees and parasitoid wasps [14,31,38] (see Figures 2 and 3). The most common outcome of duplication in *CSPs* is loss of the duplicated copy as we found in our analysis of hymenopteran *CSPs* (Figures 2 and 3). Then, there can be three different scenarios if the two duplicated copies are conserved following the gene dosage phenomenon described in the model eukaryote *Saccharomyces cerevisiae* [47] (Figure 4). In gene dosage, the two gene copies keep performing the same function as the ancestral gene and thereby introduce increased activity of the gene. Here it is a gene dosage phenomenon, i.e., the need for duplication events for sharing functions. However, duplication can also lead to restricted function or complete loss of function (Figure 4). The two events seem to have happened in the *CSP* family. Most of *CSPs* are *single-intron* genes, thus representing more restricted function [14,15,27,30,38] (also see Figures 2 and 3). Other *CSPs* are truncated unexpressed pseudogenes as found in *B. mori* and *T. castaneum* [15,38]. At a later stage of evolution, the different functions can be divided over some additional successive duplications (subfunctionalization or functional specialization of the two gene copies). Then, with one duplicated copy still performing the original function of the ancestor gene, some other new copies of the gene were subjected to mutations through or mediated via RNA editing and acquired new functions as described in Lepidoptera [13,16,17,19,20]. In particular, some specific RNA variant isoforms may have returned to the genome through or via retrotransposition to drive evolution in some groups of genes as proposed for *Bombyx CSPs* [16,20] (Figure 4).

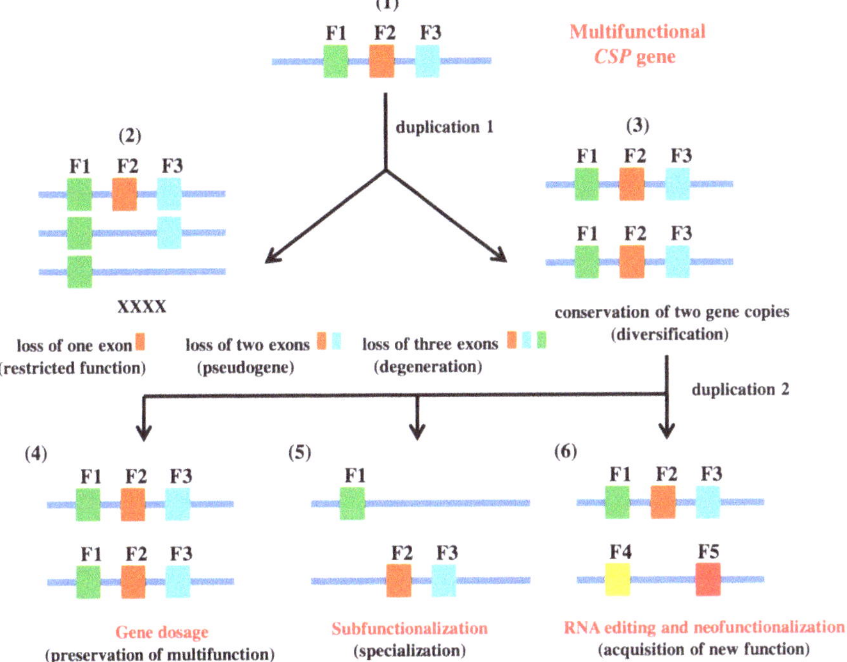

Figure 4. Evolution of *CSPs* for neofunctionalization. (**1**) At some point far back in time (Bya), the original ancestor *CSP* gene retains three functions (three exons: F1, F2 and F3). The outcome of the first early duplication (duplication 1) is a pair of tandem paralogous genes strictly identical to the original gene. (**2**) Duplication 1 can lead to loss of one exon (restricted function), loss of two exons (unexpressed pseudogene: *BmorCSP5*, *BmorCSP16*, *BmorCSP18* and *AAJJ0269A1B* [14,38]), or loss of the three exons (degeneration/loss of a complete gene copy). (**3**) Duplication 1 can also lead to the conservation of the two gene copies, increasing the original functions while providing a template for CSP diversification. Further successive genome duplications (duplication 2) allow expansion of the gene family. (**4**) Additional copies of the *CSP* gene are preserved and keep performing the same functions (F1–F3) as the ancestral gene, thus amplifying the original activity of "*CSP*". (**5**) The different functions of *CSP* are divided over specific duplicated copies, leading to CSPs with more specialized functions (subfunctionalization). (6) Other duplicates are subjected to RNA editing events and multiple specific mutations that lead the *CSP* family to acquire new genes and functions (F4, F5).

3. *CSP* Gene Expression in Response to Environmental Change

Not only the knowledge of how CSP-encoding genes have evolved, but also their tissue expression profiling are important to solve the function of the protein. For instance, CSP expression is detected during early embryonic development stages of the brine shrimp *Artemia franciscana*, clearly rejecting a function in olfaction for this protein [48]. The brine shrimp is a micro-crustacean rather known for producing cysts (dormant eggs) well adapted to harsh and critical life conditions. The CSP protein family is commonly found in many various organisms from bacteria to insects and crustaceans, including marine arthropods (that do not respond to airborne odor volatiles), and certainly they have a crucial role to play in the molecular mechanisms underlying adaptation to new environments, rather than olfaction.

Organismal adaptation to a new environment may start with very general metabolic pathways leading, for instance, to the degradation of toxic xenobiotic factors. Bacteria have no neurons and no olfactory receptors, but they are capable of chemotaxis, i.e., they can redirect their movements in the presence of chemical (amino acid or sugar) gradients [49]. Multiple *CSPs* are expressed in many various bacterial strains such as *A. baumannii*, *K. griseola*, *K. purpeofusca*, *K. CB01950*, *K. MBT66*, *E. coli* and *M. caseolyticus* [30] (Figure 3), but their role in binding solute ligands such as amino acids or sugars as well as their obvious presence in the "olfactory" hedonics of bacteria are far to be proved. Meanwhile, most bacterial species are known to readily adapt to their new environments and to develop multiple ways of multidrug chemical resistance. Therefore, studying *CSPs* may significantly help test the hypothesis that this family of genes is particularly crucial for adaptation mechanisms and evolution of cells. The accumulation of data in insect *CSPs*, in particular in moths, can now help us provide a remarkable insight into the hypothesis that the genetic plasticity in *CSPs* underlies cell fate and evolution.

The insect EST database, consisting of more than thirty thousands of mRNA sequences from *n* tissue libraries, by definition, contains information for the association of genes with tissues of origin. EST profiles in the bee *A. mellifera* do not show *CSP* gene expression restricted to the "olfactory" or "chemosensory" system. They show gene expression patterns for *CSPs* in (1) the head, (2) the brain, (3) the antennae, and 4) the whole body [39,40] (Table S3). Similarly, *CSPs* from the tiny wasp *N. vitripennis* are not expressed only at the adult stage, but they also express in larvae, prepupae and pupae [42,43]. More than 200 EST sequences are reported in the whole body of the fly *D. melanogaster* for *pebIII* and *CG9358 CSP* genes [50–52].

This is consistent with pioneer Northern blot experiments showing that moth *CSPs* are highly expressed not only in the antennae, but also in the legs, as well as in the three main parts of the insect body, head, thorax and abdomen [4–6]. The analysis of EST sequence database in the silkworm moth *B. mori* (KAIKObase) shows that the EST-cDNAs encoding BmorCSP are very abundant in antennal tissues, the compound eyes (the ocelli supply insect vision), the midgut, the ovaries, the fat body and the female pheromone gland. In the silkworm larvae, *CSPs* are found to be expressed in many various different types of tissues such as the hemocytes, the testis, the posterior silk gland, the epidermis and the maxillary galea (the sensory mouth part of the larva) [53]. Therefore, the distribution of ESTs encoding CSPs shows that these proteins are broadly expressed in early and late stages of the developing insect and absolutely never maintain a specific domain of sensory, non-sensory or neural tissues, in particular in adults.

Importantly, a more detailed gene expression study focusing on each gene in the *BmorCSP* family using real-time PCR showed unequivocally that all *CSPs* are expressed in all various tissues at the adult stage [14]. This is in agreement with the finding of wide expression of *BtabCSP1* across many different adult tissues in the whitefly *B. tabaci* and the binding of the protein to a fatty acid molecule such as C18:2 lipid, linoleic acid [15]. The molecular study from Xuan et al. comparing gene expression of all of the twenty *CSPs* from the silkworm moth *B. mori* (seventeen functional genes, three truncated pseudo-genes: *BmorCSP5*, *BmorCSP16*, *BmorCSP18*) is very important in an analysis of the role of *CSPs* in cell adaptation for three reasons: the results show that (1) none of the *BmorCSPs* is specifically expressed in one given tissue, (2) all the *BmorCSPs* are widely distributed across the insect body and

(3) about all of the seventeen functional *BmorCSP* genes show higher expression following exposure to abamectin insecticides. This strongly suggests a function in relation with immune responses, in particular in xenobiotics degradation for the whole *CSP* family (Figure 5). The three truncated genes (*BmorCSP5*, *BmorCSP16*, *BmorCSP18*) are not expressed in all tissues investigated (antennae, legs, head, pheromone gland, wings, thorax, epidermis, fat body and gut), demonstrating the loss of function after truncation of a duplicated gene [14].

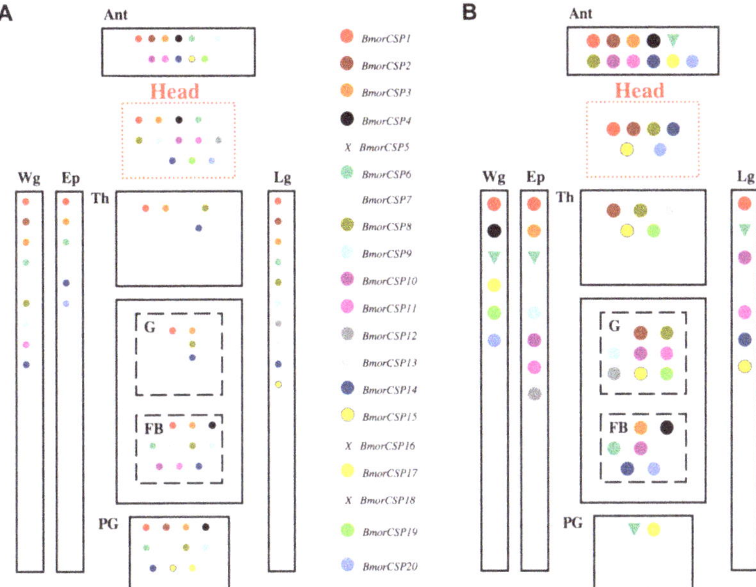

Figure 5. Tissue expression profiling of *CSPs*. *B. mori CSP* gene expression profiling under normal conditions (**A**) and following exposure to abamectin insecticide (**B**). Data are from Xuan et al. [13,14]. Specific gene expression is shown by color code. X indicates no expression for truncated genes (*BmorCSP5*, *BmorCSP16* and *BmorCSP18*). Up regulation in the expression levels of *CSP* genes is indicated by a larger circle. Down regulation in the expression levels of *CSP* gene (*BmorCSP6*) is indicated by a triangle oriented down. Ant: Antennae, Ep: Epidermis, FB: Fat Body, G: Gut, Lg: Legs, PG: Pheromone Gland, Th: Thorax, Wg: Wings.

Most *CSPs* are expressed under control conditions in a tissue-specific manner, none of the genes is consistently restricted to a common single tissue, and the expression of the whole group of *BmorCSPs* is drastically increased in a tissue-specific manner in response to a chemical stress such as the exposure to an insecticide molecule (Figure 5). It seems like a metabolic chain that enrolls most *CSPs* in the same process, i.e., the same fueling system that is essential for many various cells, organs and tissues from an organism even under normal conditions, i.e., no change of environment. Our previous study in moths shows that under no chemical or viral stress conditions, twelve to fourteen *CSP* genes are expressed in the head and peripheral organs, but their expression is never restricted to nerves or sensory tissues. About six to nine *CSP* genes are mainly expressed in the epidermis, thorax and/or the gut tract, definitely rejecting a function tuned to olfaction, chemosensing or chemotaxis. About ten to eleven *CSP* genes are mainly expressed in the pheromone gland and fat body, which are two crucial organs for mechanisms involved in lipid fatty acid uptake, metabolism, transport and trafficking (Figure 5) [14]. Free lipids and fatty acids are essential as fuel molecules for cells to regulate activities such as hormone biosynthesis, digestion, locomotion, flying, pheromone production, insecticide xenobiotic degradation and/or various immunological responses to bacterial/viral infection or host-plant poisoning. Even more

CSP genes are turned on upon abamectin insecticide exposure (Figure 5). Under severe toxic chemical stress conditions, a drastic up-regulation of a *CSP* chainwork is observed in the tissues involved in lipid metabolism and xenobiotic degradation, i.e., gut, fat body and epidermis (Figure 5).

Interestingly, among these twenty *BmorCSP* genes, only one (*BmorCSP6*) shows decreased gene expression (down-regulation) over insecticide exposure in many various tissues such as the antennae, the pheromone gland, legs and wings as well as epidermis (see Figure 5) [14]. This may suggest that BmorCSP6 has a very different function than the other BmorCSPs. The expression of *BmorCSP6* in bacteria and in many various insect tissues from epidermis to pheromone gland strongly argues that olfaction and pheromone production are not BmorCSP6 primarily functions. Most surprising and interesting fact is that, alike BmorCSP2, BmorCSP6 exists identically in insects and bacteria [30]. The function of these highly conserved protein sequences is unknown. They may play a conserved role in mechanosensing of droplets or surfaces, i.e., the process that often uses obstruction of flagellum rotation to trigger adhesion, surface-associated movement, biofilm formation and/or bacterial virulence [54,55]. This could explain their contribution to insect epidermal cells, glands, neurons and bacteria. So it could be that one chainwork of *CSPs* is turned off upon exposure to insecticide or bactericide, while another *CSP* chainwork is activated to degrade or expel the infectious toxic agent in a sex or strain-specific manner (Figure 5) [14].

4. CSPs for Lipid- and Fatty Acid- Mediated Pathways

Also interestingly, numerous *CSP* genes are used in the nervous system of the silkworm moth, *B. mori*. Under normal conditions, nearly all *CSPs* are expressed not only in the head, but also in moth peripheral organs such as the antennae, the wings and the legs (Figure 5) [14].

A similar observation was made by Liu et al. (2016) related to this finding, *CSP* expression throughout the whole body, but mainly in the head and the peripheral organs [15]. With this study, we give two main points for assessment of CSP function in insects: (1) *CSP* is up-regulated by insecticide, and (2) the protein binds specifically to linoleic acid, strongly arguing for a role in fatty acid- and lipid-mediated pathways for adaptation, signaling and immune defense [15]. Most importantly for our present analysis of function is that annotated ESTs in the honeybee show enriched expression of *CSP* in the head, particularly in the brain (see Table S3), strongly suggesting that CSPs have complex actions not only in the insect immune system, but also in the central nervous system and virtually all of the body's organs, including antennae, wings and legs.

Based on these results, we propose that CSPs play a key role in activating the omega6 fatty acid pathway, which is necessary to produce diacylglycerol (DAG) that will in turn activate phospholipase kinase C and phosphorylation of many various different proteins (Figure 6) [56]. DAG-mediated protein phosphorylation is an essential requirement not only of neuron depolarization/repolarization (sodium/potassium channels), signal transduction (transmembrane receptor) and/or muscle contraction (myosin motor protein), but also of the activation of lipid biosynthetic and degradative enzymes such as cytochrome oxidases (CYPs) and delta (Δ)-desaturases (Figure 6). These enzymes both regulated by phosphorylation/dephosphorylation processes are essential for xenobiotic degradation, storage of fatty acids and pheromone production [57–62]. Accordingly, they represent very important molecular elements for organismal adaptation and evolution.

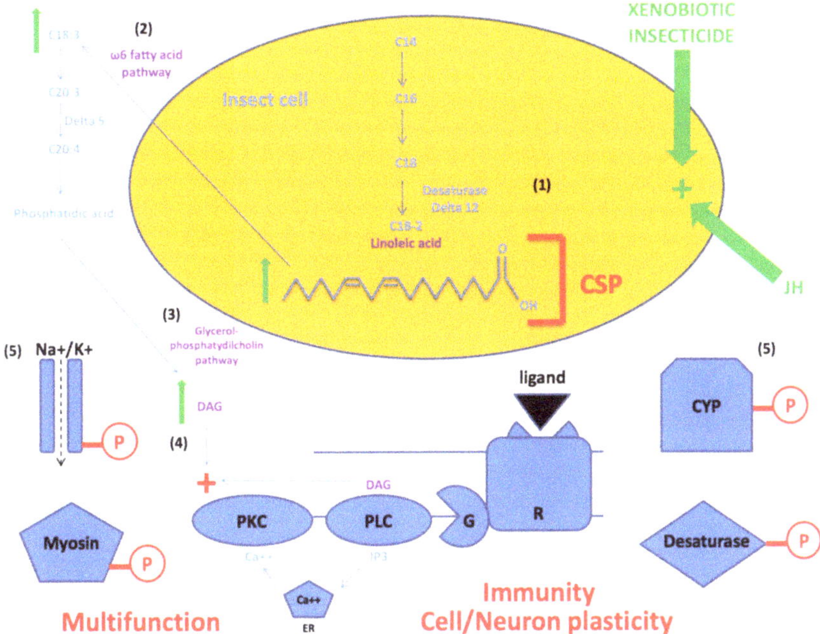

Figure 6. Conjectural model of the role of CSP in activation of linoleic acid/omega6 fatty acid-diacylglycerol pathway upon xenobiotic insecticide or juvenile hormone exposure. (**1**) Some insect cells have the ability to synthesize linoleic acid (C18:2) *de novo* using fourteen-eighteen carbons-fatty acids (C14–C18) and specific desaturases. (**2**) C18:2 is fuel molecule for omega6 fatty acid pathways. Molecules such as arachidonyl-CoA ((Z5,Z8,Z11,Z14)-Icosatetraenoyl-CoA or C20:4) are products of a Δ5 desaturase reaction from eicosatrienoyl-CoA (C20:3) as a direct substrate. (**3**) Synthesis of these fatty acid metabolites leads to phosphatidic acid and therefore to the formation of diacylglycerol (DAG) through the biosynthetic pathway of glycerol-phosphatydilcholins. (**4**) DAG is a relay molecule in intracellular cascades activated by the binding of regulatory chemical ligand (labelled by a black triangle) to G-protein coupled receptor. This triggers the formation of inositol 1,4,5-triphosphate (IP3) and DAG by PLC (phospholipase C). In turn, IP3 releases the calcium ions (Ca^{++}) from intracellular stocks in the endoplasmic reticulum (ER). (**5**) DAG (with Ca^{++}) activates (**+**) protein kinase C (PKC), which in turn induces specific cellular responses by phosporylating a particular set of cellular proteins (ion channels, myosin, cytochrome P450, desaturase enzymes, etc.). Applying xenobiotic insecticide and/or juvenile hormone (JH) activates (**+**) the DAG pathway and thereby protein phosphorylation (red symbol P) via increased concentrations of C18 and C20 fatty acids on cell growth performance and/or immune response of various tissues, organs and organ systems. The green arrow means that the concentration of C18:2, C18:3 and DAG increases with increasing concentration of xenobiotic insecticides and/or JH. The central role for CSP in Δ12-fatty acid pathway associated with transport of C18:2 for multifunction, immunity, cell development, tissue growth and neuronal plasticity is shown in red.

In insects, it has been shown that the DAG-phospholipase C (PLC) pathway can be regulated by juvenile hormone (JH) [63], while exposure to pyrethroid insecticide can stimulate protein phosphorylation activity in the brain of mammals [64]. This suggests that insecticide exposure can stimulate the DAG-PLC pathway in many various tissues of the insect body, not only by a stimulatory effect on the production of C18-linoleic acid (LA) and lipid omega6 fatty acids production, but also via an indirect effect on JH release [65–67]. A plethora of pleiotropy across CSPs and binding protein families seems to be necessary to recognize a multitude of targets and phosphorylation sites in a huge variety of complex cell-cell and intracellular signaling pathways in many diverse organisms.

Interestingly, microbes are known to produce LA and to carry a gene related to JH [68,69]. Both LA and JH are known to be crucial for many cell functions in worms and arthropods, particularly in growth, developmental, reproductive and innate immune systems [70–72]. In insects, JH is a sesquiterpenoid hormone produced by the *corpora allata* and is present throughout nymphal, larval and adult life. Most insect species produce only one JH-type (type III), but only butterflies (and moths) produce other JH types such as JH-0, JH-I and JH-II. Flies produce the form JHB3 (JH-III bisepoxyde) [73]. For LA, it has long been a debate about synthesis of C18:2$^{\Delta 9,12}$ lipids in insects as most species were thought to be lacking Δ12 desaturases, the enzymes capable of inserting a double bound at the Δ12-position. In fact, it is clear now that insects such as ants, bees, cockroaches (*Periplaneta*), crickets, moths, termites and wasps can synthesize LA *de novo* [74]. It is well known that oleic acid and LA are "necromones", pheromones given off by a dead organism as described in ants, bees, cockroaches and crickets [75]. Oleic acid is also known as a main precursor molecule of LA and male sex pheromone in the parasitoid wasp *N. vitripennis* [76]. Similarly, LA is a precursor of sex pheromone compound in moth species such as Bombycidae, Crambidae and arctiid moths [77–79]. Finally, LA and C18 fatty acids are known to play a key role in moth development as demonstrated in the crambidae species, *Ostrinia nubilalis* [80]. Therefore, the interactions of CSPs and LA/linolenic acid on fatty acid pathways would be crucial to regulate many various physiological systems, including pheromone biosynthesis, growth, development, tissue regeneration and/or toxin/insecticide immune responses in many various species from bacteria to all various groups of insects.

Accordingly, an adequate number of CSPs may be important to interact with the various intermediary molecules of the LA/fatty acid pathway. EFN87902/EFN72587, NV16109 and AmelGB10389 would have the function to transport LA in hymenoptera because they clearly group together with BtabCSP1 in our evolutionary analysis of insect CSPs (see Figure 3). The bee protein AmelASP3c is highly expressed in the antennae (in sensilla trichodea B and sensilla basiconica), but it is absolutely not restricted to antennal sensilla. It is also found on wings and legs, suggesting a very much more general function than queen pheromone recognition for this CSP protein [81,82]. The tissue distribution of CSP-EST sequences in the honeybee *A. mellifera* confirms that *ASP3c* is not specifically expressed in a peripheral organ such as the antennae. We find that the part of the insect body that expressed *ASP3c* gene the most is the head, and the main organ for ASP3c is the brain (Table S3). So, we propose that ASP3c transports fatty acid lipids instead of pheromones for instance for process, growth, development and/or regeneration in neurons and other cell types harbored by the central nervous system in the honeybee.

Similarly, in situ hybridization to RNA in antennal tissue section from *D. melanogaster* shows that CSP (DmelOSD) associates with the sacculus (involved in hygrosensing) and patches of sensilla coeloconica distributed on various parts of the fly antennae [83,84]. DmelOSD homologs are also found in the hemolymph in response to microbial inoculation [46]. Therefore, it is clearly shown that many various types of sensilla and tissues as well as the circulatory fluid bathing these tissues possess the same CSPs. Such expression profiling in Hymenoptera and Diptera is in agreement with a role of CSPs in the transport of lipids, which are essential for cell survival and adaptation. The gene *CG9358* (related to *AmelGB13325* and *NV16108*) is under the control of embryo and tissue developmental factors and governed by transcription factors involved in circadian rythms [85–87]. So, it is worth noting that flies that express only four *CSPs* cannot synthesize LA de novo and therefore require a dietary resource of C18:2 lipid [12,37,88]. On the basis of numerous observations, we propose that the inability of flies to synthesize LA and LA derivatives is caused by lack of specific enzymes, i.e., Δ12 desaturases, and a specific group of small transport proteins of the CSP family (Group V, see Figure 3), as an example of the importance of *CSPs* in cell evolution.

5. Genetic Editing of CSPs for Insecticide Resistance

Interestingly, it is also worth noting that the *CSP* family is subjected to RNA editing and that this RNA editing may significantly increase expression and activity of the protein (see Figure 1) [13,16,17,19,31].

Edited versions or mutations are not limited to A-to-I and/or C-to-U conversion, but protein diversity and multifunction in CSPs seems to be brought on by many other mechanisms, including insertion of specific amino acid motifs and residues at the protein level [13,16,17,19,31] (see Figure 1). Insertion of amino acids such as Glycine near Cysteine at key position on the protein structure may modify the profiling of alpha-helices, which are essential components of the protein-fatty acid lipid interaction [7–11,13,16,17,19,20]. The data obtained in moths corroborate the hypothesis that RNA editing compensates for a small genome size and/or for the decreased diversity of *CSP* genes, as reported here for honeybees and parasitoid wasps. RNA editing in *CSPs* seems to be crucial for lipid transport and thereby cell type diversity.

In these insect species such as lice, bees and wasps with only six to eight *CSPs*, only RNA and/or protein editing mechanisms could allow them to use CSPs for the transport of a high diversity of lipid-ligands as proposed for moths [13–17,30,31] (see Figures 1–6). In beetles, a huge amount of RNA variants are found for *AAJJ0012I* and *AAJJ0283B* [38]. In locusts, a high number of copies of genes (> fifty) are used for differential expression pattern of *CSPs* in relation with phase change [89,90]. However, some of the RNA clones identified in *Locusta migratoria* adults indicated the presence of subtle nucleotide replacements (A-to-G, A-to-C, C-to-A and U deletion) between some specific *CSP* sequences, suggesting the occurrence of RNA editing in *CSPs* not only in moths and beetles, but also in locusts and grasshoppers in the Acrididae family [4]. The American cockroach (*P. americana*) shows numerous variant N-terminal sequences for CSP proteins, similarly to *Bombyx, Tribolium* and *Locusta*. So, this genetic regulation or plasticity of *CSPs* through RNA editing also occurs in the order Blattodea [2,5]. Mutation is also described in whiteflies where *CSP* mutations appear to be biotype-specific [15]. Many mutant peptide fragments sequenced in *Bombyx* are very similar to CSPs from Hemipteran or Dipteran species, strongly suggesting that RNA editing of *CSP* is important for many insect species [13,20].

Here, we report about the importance of RNA editing in *CSPs* from groups of social insects such as the honeybee. Analyzing EST sequences from GenBank [91], we find a high number of mutations (RNA editing) in the order Hymenoptera, particularly in the bee brain (Table S3).

Analyzing nucleotide sequences encoding CSP in *A. mellifera* using blastn algorithm for GenBank EST sequences in FlyBase shows numerous subtle nucleotide switches such as A-to-G and U-to-C at least for three *CSPs* from the honeybee: *AmelASP3c*, *GB17875* and *GB19453* (Table S3). Such a high number of mutations in the bee brain suggest the importance of RNA editing in *CSPs* for the insect central nervous system (Table S3). Editing of *CSPs* (A-to-G, U-to-C, C deletion, U-to-G, G-to-U, U-to-A, A-to-C, A-to-G, G-to-C and C-to-G) in the bee brain (Table S3) may be linked to task performance and social behavior [92,93].

Subsequently, we propose that RNA editing as well as rapid evolution and positive selection in *CSP* duplicates (increasing the number of *CSP* genes) has largely contributed to the development of new protein functions, resulting in high insecticide resistance capacities, for instance, particularly in insect species such as beetles, moths and whiteflies [13–15,38]. The beetle has developed resistance to more than fifty different chemicals belonging to all major insecticide chemicals [94,95]. Correlatively, it expresses a number of about nineteen *CSP* genes [38]. Moth larvae that can develop very fast insecticide resistance even to new chemicals retain about twenty *CSP* genes as described in the silkworm *B. mori* [14,96,97]. In the whitefly *B. tabaci*, some biotype-specific variations exist within *CSPs*, which could underlie such a high insecticide resistance capacity observed in whiteflies of Q-biotype, in particular for neonicotinoid molecules [15,98,99]. So not only detoxification genes, but also the number of *CSPs* and/or *CSP*-RNA variants as well as their ability to bind to specific lipids, LA and other fatty acids, as well as xenobiotic compounds, may be crucial for diverse insect species, strains or biotypes to develop a high resistance capacity to chemical insecticide molecules.

Insect CSPs are crucial in chemical communication by recognizing environmental chemical stressors [14,15]. The RNA editing of *CSP* genes is not the only way to be involved in insecticide resistance in insects [13,16,17]. Besides expression and mutation, there are examples of multifunction to explain how *CSPs* may play a central role in insecticide resistance. In moths, *CSPs* respond to avermectins [14]. In whiteflies, CSPs are lipid carriers and xenobiotic transfer proteins [15]. All RNA editing in *CSPs* may be associated with insecticide resistance for the transport of many different types of ligands and/or activation of a variety of degrading enzymes such as cytochromes P450 (CYPs) and JH esterases [14,15]. *Bombyx* CSP-RNA is characterized by high mutation rates in many various sensory and non-sensory tissues, including the pheromone gland [13,16,17]. In the flour beetle *T. castaneum*, CSP-RNA variants are mainly found in the hindgut and Malpighian tubules in response to insecticide [38]. RNA editing is also crucial to mediate insecticide (ivermectin) resistance through specific point mutations in GABA receptors (resistant to dieldrin) [100]. Similarly, RNA editing in sodium channel in mosquito plays a role in pyrethroid resistance [101], however, another study showed that RNA and genomic DNA sequences from the same *Aedes aegypti* individual did not support the involvement of RNA editing in permethrin resistance [102]. Therefore, the importance of RNA editing in insecticide resistance may depend on chemical families, insecticide structures and insecticidal properties or modes of action. RNA editing in *CSPs* is an extremely important component of resistance to avermectins known to block the transmission of electrical signals in insect nerve and muscle cells by targeting glutamate-gated chloride channels, and neonicotinods, which target nicotinic acetylcholine receptors (nAChRs) [13–17]. More work needs to be performed to check whether all RNA editing in mechanisms and molecular targets of avermectin and neonicotinoid pesticides are associated with insecticide resistance.

6. Genetic Plasticity of *CSPs* for Neuroplasticity

So, the bees being so depauperate genetically, lacking many detoxification enzymes and CSPs, become very sensitive to the toxicity of many various foreign chemicals [103]. This does not, however, exclude the possibility that other CSPs or groups of CSPs may be essential for the bees to accomplish special feats such as odor memorization and specific social behavior. The high expression levels of CSPs and CSP variants that we have detected in the nervous system of the honeybee, particularly in the brain (see Table S3), suggest a possible role of CSPs in learning and memorization processes. A possible role of CSPs in neuroplasticity is also suggested by gene knockout experiments. *CSP* (*AmelGB10389* on LG1 chromosome) has been knocked out in bees, resulting in an archaic development of the brain [104].

When a cell differentiates or acquires a defined specialized function, it is supposed to undertake major changes in its size, shape, protein synthesis, metabolic activity, and overall function which at the end will serve a defined tissue or organ. Despite their different shapes, colors and functions, all various tissues or organs, even in most complex multi-cellular organisms such as humans, come from the same basic common totipotent cell, i.e., an immature stem cell capable of giving rise to any cell type from an embryo or an undifferentiated cell that can renew itself and can differentiate to provide any specialized cells types of a given tissue or organ in an organism at a certain point in time [105–108]. Our finding in insects indicating such a huge diversity in base mutations and protein changes at the level of CSPs may bring an answer not only about the high capacity of insects for chemical resistance, but also about the basis for the development of stem cells as well as for structural and functional reactions of neurons.

Neuronal plasticity is reported in arthropods and insects as various responses involving any change in the brain from dendrite regeneration, axon sprouting and synapse formation, resulting in specific behavioral adaptations [109]. It could be that the multi-function of CSPs in carrying all sorts of lipids and adhering to all sorts of surfaces plays a key role in the mushroom body neuropiles, i.e., in olfactory learning and memorization processes, in particular in adult social insects such as the honeybee and long-lived migrant species of moths such as the black cutworm moth *Agrotis ipsilon* [110]. In long-lived species of moths, it has been shown that JH known to exert pleiotropic functions during the whole insect life cycle, controlling many various physiological systems from metamorphosis, tissue development and pheromone activities, is also essential for peripheral and central nervous processing of sex pheromone and/or plant odor [110–113]. Pleiotropic proteins such as CSPs and pleiotropic hormones such as JH may interact with each other to govern the switches observed in the brain responses to odorant signals. Controlling LA pathways (see Figure 6), both CSP and JH may allow differential processing of pheromone and plant odor, i.e., activation or transient blockade of specific integrative centers in the brain [112]. This needs to be elucidated by searching for the ability of CSPs to interact with JH and/or to locate precisely the site of CSP expression not only in the brain structure, but also in the bee or moth neuron.

Expression of *CSPs* in the neural system of insects to control DAG and protein phosphorylation may be an example of neofunctionalization of this protein gene family for neuroplasticity, neurogenesis, synaptogenesis, the formation of new synapses and generation of new neuron connections (see Figure 6). So far, we can only discuss abundant pleiotropy in CSPs for insect defense and lipid metabolism [13–19]. Pleiotropy (functional plasticity or multi-function) of CSPs is demonstrated by the study of Liu et al. in the whitefly *B. tabaci*, where CSP1 is involved in the response against thiametoxam by interacting with LA, while two other CSPs rather involve in the transport of bark plant phenolic chemicals such as cinnamaldehyde and derivatives [15]. Cinnamaldehyde and cinnamon leaf oil are known to retain the ability to kill bacteria, fungi, mosquito larvae and many insects on contact as well as to act as a strong repellent long afterwards [114]. Therefore, cinnamon oil seems to represent a very ancient system that plants have developed for defense against insects and microbial pathogens. Plants have been interacting with herbivorous insects and bacterial fauna for hundreds of millions of years. In turn, herbivorous insects (and bacteria) have certainly developed their own defense system to counteract the panoply of poisonous chemicals released by the plant for My(Million years). Liu et al. have demonstrated that CSPs are essential for cell defense through the binding of lipids and xenobiotics [15]. Therefore, lipid transport and the sequestration of toxic xenobiotic chemicals may represent some ancestral CSP functions, i.e., used by bacteria and early eukaryotes to grow and adapt to the natural environment. Later, new CSP functions may have been crucial for the appearance and development of the nervous system, including not only the formation of many types of brain cells, but also neural plasticity.

A lack of variation for genetic plasticity, RNA editing and/or protein recoding would have led to a lack of evolutionary perspective for adaptive capacity in all diverse organismal associations such as the moth and the green plant or the bee and the flower. The complex system of pleiotropic genes such as *CSPs* enrolled not only in lipid and FA biosynthesis, incorporation, transport and metabolism, but also in immunity, tissue growth and neuroplasticity is certainly a big part of the most ancient evolutionary components of the cellular system of living organisms in an environment that is constantly changing.

7. Conclusions and Future Research

In this review, we do not give justice to the eluding nature of the CSP protein family, but address all the known aspects of this protein gene family: the post-transcriptional modification of the genes encoding chemosensory proteins, the genomic organization of *CSPs* as described here in honeybee and jewel wasp, the phylogenetic distribution of CSP sequences from insects/arthropods and bacteria, their gene expression profiling and tissue-distribution, and their multifunctionality, as well as their role in lipid fatty acid pathways for various physiological systems, including mainly insect defense and insecticide resistance. Then, we attempt to engage in an understanding of their neofunctionalization or ability to interact with lipids and fatty acids for neuroplasticity.

Although the biochemical mechanisms of CSPs in the resistance against insecticide has not been fully investigated, a role of CSPs at different levels of the insect immunological defense is strongly supported by the ability of whitefly CSP1s to interact with lipids, while whitefly CSP2s and CSP3s have the ability to interact directly with specific xenobiotic compounds such as cinnamaldehydes from plant oils [15].

An issue for debate about a common role of the CSP workchain for neuroplasticity is that most CSPs in the silkworm are enrolled in the nervous system upon normal conditions [14], and that most CSPs in the honeybee are expressed in the brain (this study). It remains to be found if they are involved in the same process from bacteria to insects, if they all have the same function that many organisms from bacteria to insects use, or if some of these CSPs were subjected to specific mutations and acquired a more specialized new function, prior to the birth and development of neuronal cells and/or specific behavioral traits, including those of social insects such as the honey bees and those of migrant species such as the black cutworms.

In these species, research should be made for RNA/protein mutation on a specific tissue such as the brain and analysis of functional properties, particularly in the insect neuropile where a dense network of nerve fibers, their branches and synapses, together with glial filaments rebuild and reorganize specific synaptic connections, especially in response to learning, memorization and/or brain tissue injury. Pleiotropic CSPs capable of carrying fuel molecules such as lipids and fatty acids might be crucial in these processes of development and neural tissue regeneration.

In addition, deeper research should concern CSPs, immune cells and/or cells exposed to a panoply of antigenic substances. Human thymus or insect hemocytes can express a prominent diversity of proteins and protein variants in response to infection or environmental contamination, and yet adapts to new conditions and sustains development as well as natural evolution. Apparently, considering the genetic plasticity, RNA editing and true functional pleiotropy characterizing this gene family, *CSPs* could potentially bring an answer for stem cell research, phenotypic evolution and critical thinking in questions of neuroscience. Firstly, because all cells in the body, beginning with the fertilized egg, contain the same DNA, how do the different cell types come to be so different and different enough to yield such a high diversity of tissues or organs, each characterized by a specific specialized function? Secondly, how can the insect brain switch on/off its responses to specific odor signals depending on the environment?

Far beyond the DNA structure, it is well established now that post-transcriptional events such as alternative splicing and RNA editing are able to subtly modify proteins to diversify their structures for multi-function. In particular, new mechanisms to be found in the expression of CSPs may serve to explain neuroplasticity in the nervous system, the diversity of cellular responses in the immune system, and fate as well as transformation in the stem cells of a newborn organism [19,31,115]. In this review of multiple genetic events, using *CSPs* as a model study, we discuss how RNA editing and activation of lipid fatty acid pathways can contribute to specific innate and adaptive immune responses and/or to specific neurobiological development (brain–immune interactions) in parasitoid wasps and social insects such as the honey bees. RNA editing is probably required to circumvent a rather limited repertoire of *CSP* genes as found for parasitoids and bees. We find only six and eight *CSP* genes in the honeybee *A. mellifera* and the solitary parasitoid emerald jewel wasp *N. vitripennis*, respectively. Gene structure and intron boundary show gene duplication, but our phylogenetic tree analysis shows a distinct evolutionary route between honeybee and pteromalid parasitoid wasp *CSPs*. We report here that a particular group of "ancient" *CSPs* is closely related to metabolic *CSPs* from bacteria and aquatic species of arthropods, perhaps suggesting that *CSPs* are the products of a duplication that took place Bya in the most ancient (Archaeal) organismal lineage and it is very likely that this duplication happened to be crucial for the adaptation of Archaeal cells.

Further duplications might have happened to promote adaptation and evolution of prokaryote and eukaryote cells in diverse environments [116]. Therefore, genetic plasticity (gene duplication and RNA editing) in *CSPs* should be investigated not only in the neural stem cells of the insect brain, but probably also in the filamentous bacterial cells to uncover cell–cell adhesion and interaction mechanisms. This would be an essential prerequisite to understand neuroplasticity, tissue differentiation, organ development, cell proliferation, bacterial infection, virulence and immune defense. Genetic editing or RNA plasticity in *CSPs* is a very new and promising subject to allow for a better understanding of the role of small soluble binding protein carriers in insects and bacteria to be explored by both entomological and medical healthcare industries and, most likely, of evolutionary processes that gave rise to life diversity at every level of biological organization [117].

Supplementary Materials: The following are available online at http://www.mdpi.com/2073-4425/11/4/413/s1, Figure S1: Genomic organization of *B. mori* and *T. molitor CSPs* [14,41]. The red arrow shows the position of *double-intron* genes (outside the main group of *CSPs*). Table S1: *CSP* genes identified by in silico analaysis of honeybee *A. mellifera* database. Table S2: *CSP* genes identified by in silico analaysis of jewel wasp *N. vitripennis* database. Table S3: Tissue distribution of CSP-EST sequences in honeybee *Apis mellifera* (*carnica*).

Author Contributions: Conceptualization, J.F.P.; methodology, J.-F.P.; software, B.R., P.A., B.O., J.-F.P.; validation, J.-F.P.; formal analysis, J.-F.P.; investigation, G.L., N.X., B.R., P.A., B.O., J.-F.P.; resources, G.L., N.X., J.-F.P.; data curation, J.-F.P.; writing-original draft preparation, J.-F.P.; writing-review and editing, J.-F.P.; visualization, G.L., J.-F.P.; supervision, J.-F.P.; project administration, J.-F.P.; funding acquisition, G.L., J.-F.P.; All authors have read and agreed to the published version of the manuscript.

Funding: This research was funded by Natural Sciences Foundation of Shandong Province, Overseas Talent, Taishan Scholar, Agricultural Scientific and Technological Innovation Project of SAAS and Key Research and Development Program of Shandong Province, grant numbers ZR2011CM046, No. tshw20091015, CXGC2016A11 and 2019GSF107085.

Conflicts of Interest: The authors declare no conflicts of interest.

References

1. Nomura, A.; Kawasaki, K.; Kubo, T.; Natori, S. Purification and localization of p10, a novel protein that increases in nymphal regenerating legs of *Periplaneta americana* (*American cockroach*). *Int. J. Dev. Biol.* **1992**, *36*, 391–398. [PubMed]
2. Picimbon, J.F.; Leal, W.S. Olfactory soluble proteins of cockroaches. *Insect Biochem. Mol. Biol.* **1999**, *30*, 973–978. [CrossRef]
3. Angeli, S.; Ceron, F.; Scaloni, A.; Monti, M.; Monteforti, G.; Minnocci, A.; Petacchi, R.; Pelosi, P. Purification, structural characterization, cloning and immunocytochemical localization of chemoreception proteins from *Schistocerca gregaria*. *Eur. J. Biochem.* **1999**, *262*, 745–754. [CrossRef] [PubMed]
4. Picimbon, J.F.; Dietrich, D.; Breer, H.; Krieger, J. Chemosensory proteins of *Locusta migratoria* (Orthoptera : Acrididae). *Insect Biochem. Mol. Biol.* **2000**, *30*, 233–241. [CrossRef]
5. Picimbon, J.F.; Dietrich, K.; Angeli, S.; Scaloni, A.; Krieger, J.; Breer, H.; Pelosi, P. Purification and molecular cloning of chemosensory proteins from *Bombyx mori*. *Arch. Insect Biochem. Physiol.* **2000**, *44*, 120–129. [CrossRef]
6. Picimbon, J.F.; Dietrich, K.; Krieger, J.; Breer, H. Identity and expression pattern of chemosensory proteins in *Heliothis virescens* (*Lepidoptera, Noctuidae*). *Insect Biochem. Mol. Biol.* **2001**, *31*, 1173–1181. [CrossRef]
7. Lartigue, A.; Campanacci, V.; Roussel, A.; Larsson, A.M.; Jones, T.A.; Tegoni, M.; Cambillau, C. X-ray structure and ligand binding study of a moth chemosensory protein. *J. Biol. Chem.* **2002**, *277*, 32094–32098. [CrossRef]
8. Campanacci, V.; Lartigue, A.; Hällberg, B.M.; Jones, A.; Giudici-Orticoni, M.T.; Tegoni, M.; Cambillau, C. Moth chemosensory protein exhibits drastic conformational changes and cooperativity on ligand binding. *Proc. Natl. Acad. Sci. USA* **2003**, *100*, 5069–5074. [CrossRef]
9. Jansen, S.; Zídek, L.; Löfstedt, C.; Picimbon, J.F.; Sklenar, V. 1H, 13C, and 15N resonance assignment of *Bombyx mori* chemosensory protein 1 (BmorCSP1). *J. Biomol. NMR* **2006**, *36*, 47. [CrossRef]
10. Jansen, S.; Chmelik, J.; Zídek, L.; Padrta, P.; Novak, P.; Zdrahal, Z.; Picimbon, J.F.; Löfstedt, C.; Sklenar, V. Structure of *Bombyx mori* Chemosensory Protein 1 in solution. *Arch. Insect Biochem. Physiol.* **2007**, *66*, 135–145. [CrossRef]
11. Tomaselli, S.; Crescenzi, O.; Sanfelice, D.; Ab, E.; Wechselberger, R.; Angeli, S.; Scaloni, A.; Boelens, R.; Tancredi, T.; Pelosi, P.; et al. Solution structure of a chemosensory protein from the desert locust *Schistocerca gregaria*. *Biochemistry* **2006**, *45*, 1606–1613. [CrossRef] [PubMed]
12. Picimbon, J.F. Biochemistry and Evolution of CSP and OBP Proteins. In *Insect Pheromone Biochemistry and Molecular Biology—The Biosynthesis and Detection of Pheromones and Plant Volatiles*; Blomquist, J.G., Vogt, R.G., Eds.; Elsevier Academic Press: London, UK; San Diego, CA, USA, 2003; pp. 539–566.
13. Xuan, N.; Bu, X.; Liu, Y.Y.; Yang, X.; Liu, G.X.; Fan, Z.X.; Bi, Y.P.; Yang, L.Q.; Lou, Q.N.; Rajashekar, B.; et al. Molecular evidence of RNA editing in *Bombyx* chemosensory protein family. *PLoS ONE* **2014**, *9*, e86932. [CrossRef] [PubMed]
14. Xuan, N.; Guo, X.; Xie, H.Y.; Lou, Q.N.; Bo, L.X.; Liu, G.X.; Picimbon, J.F. Increased expression of CSP and CYP genes in adult silkworm females exposed to avermectins. *Insect Sci.* **2015**, *22*, 203–219. [CrossRef] [PubMed]

15. Liu, G.X.; Ma, H.M.; Xie, H.Y.; Xuan, N.; Xia, G.; Fan, Z.X.; Rajashekar, B.; Arnaud, P.; Offmann, B.; Picimbon, J.F. Biotype characterization, developmental profiling, insecticide response and binding property of *Bemisia tabaci* chemosensory proteins: Role of CSP in insect defense. *PLoS ONE* **2016**, *11*, e0154706. [CrossRef] [PubMed]
16. Xuan, N.; Rajashekar, B.; Picimbon, J.F. DNA and RNA-dependent polymerization in editing of *Bombyx* chemosensory protein (CSP) gene family. *Agric. Gene* **2019**, *12*, 100087. [CrossRef]
17. Xuan, X.; Rajashekar, B.; Kasvandik, S.; Picimbon, J.F. Structural components of chemosensory protein mutations in the silkworm moth, *Bombyx mori*. *Agric. Gene* **2016**, *2*, 53–58. [CrossRef]
18. Waterhouse, A.; Bertoni, M.; Bienert, S.; Studer, G.; Taurellio, G.; Gumienny, R.; Heer, F.T.; de Beer, T.A.P.; Rempfer, C.; Bordoli, L.; et al. SWISS-MODEL: Homology modelling of protein structures and complexes. *Nucleic Acids Res.* **2018**, *46*, W296–W303. [CrossRef]
19. Picimbon, J.F. A new view of genetic mutations. *Australas. Med. J.* **2017**, *10*, 701–715. [CrossRef]
20. Picimbon, J.F. Evolution of Protein Physical Structures in Insect Chemosensory Systems. In *Olfactory Concepts of Insect Control-Alternative to Insecticides*; Springer Nature Switzerland: Basel, Switzerland, 2019; Volume 2, pp. 231–263.
21. Liu, Z.; Song, W.; Dong, K. Persistent tetrodotoxin-sensitive sodium current resulting from U-to-C RNA editing of an insect sodium channel. *Proc. Natl. Acad. Sci. USA* **2004**, *101*, 11862–11867. [CrossRef]
22. Song, W.; Liu, Z.; Tan, J.; Nomura, Y.; Dong, K. RNA editing generates tissue-specific sodium channels with distinct gating properties. *J. Biol. Chem.* **2004**, *279*, 32554–32561. [CrossRef]
23. Black, D.L. Splicing in the inner ear: A familiar tune, but what are the instruments? *Neuron* **1998**, *20*, 165–168. [CrossRef]
24. Neves, G.; Zucker, J.; Daly, M.; Chess, A. Stochastic yet biased expression of multiple Dscam splice variants by individual cells. *Nat. Genet.* **2004**, *36*, 240–246. [CrossRef] [PubMed]
25. Nishikura, K. A-to-I editing of coding and non-coding RNAs by ADARs. *Nat. Rev. Mol. Cell Biol.* **2016**, *17*, 83–96. [CrossRef] [PubMed]
26. Wooldridge, L.; Ekeruche-Makinde, J.; van den Berg, H.A.; Skowera, A.; Miles, J.J.; Tan, M.P.; Dolton, G.; Clement, M.; Llewellyn-Lacey, S.; Price, D.A.; et al. A single autoimmune T cell receptor recognizes more than a million different peptides. *J. Biol. Chem.* **2012**, *287*, 1168–1177. [CrossRef] [PubMed]
27. Bushdid, C.; Magnasco, M.O.; Vosshall, L.B.; Keller, A. Humans can discriminate more than 1 trillion olfactory stimuli. *Science* **2014**, *343*, 1370–1372. [CrossRef]
28. Liu, G.X.; Ma, H.M.; Xie, H.Y.; Xuan, N.; Picimbon, J.F. Sequence variation of *Bemisia tabaci* Chemosensory protein 2 in cryptic species B and Q: New DNA markers for whitefly recognition. *Gene* **2016**, *576*, 284–291. [CrossRef]
29. Vizueta, J.; Frias-Lopez, C.; Macias-Hernandez, N.; Arnedo, M.A.; Sanchez-Gracia, A.; Rozas, J. Evolution of chemosensory gene families in arthropods: Insight from the first inclusive comparative transcriptome analysis across spider appendages. *Genome Biol. Evol.* **2017**, *9*, 178–196. [CrossRef]
30. Liu, G.X.; Picimbon, J.F. Bacterial origin of chemosensory odor-binding proteins. *Gene Transl. Bioinform.* **2017**, *3*, e1548. [CrossRef]
31. Liu, G.X.; Yue, S.; Rajashekar, B.; Picimbon, J.F. Expression of chemosensory protein (CSP) structures in *Pediculus humanus corporis* and *Acinetobacter baumannii*. *SOJ Microbiol. Infect. Dis.* **2019**, *7*, 1–17.
32. Siepel, A.; Bejerano, G.; Pedersen, J.S.; Hinrichs, A.S.; Hou, M.; Rosenbloom, K.; Clawson, H.; Spieth, J.; Hillier, L.D.W.; Richards, S.; et al. Evolutionary conserved elements in vertebrate, insect, worm and yeast genomes. *Genome Res.* **2005**, *15*, 1034–1050. [CrossRef]
33. Isenbarger, T.A.; Carr, C.E.; Johnson, S.S.; Finney, M.; Church, G.M.; Gilbert, W.; Zuber, M.T.; Ruvkun, G. The most conserved genome segments for life detection on earth and other planets. *Orig. Life Evol. Biosph.* **2008**, *38*, 517–533. [CrossRef] [PubMed]
34. Zhu, J.; Wang, G.; Pelosi, P. Plant transcriptomes reveal hidden guests. *Biochem. Biophys. Res. Commun.* **2016**, *474*, 497–502. [CrossRef] [PubMed]
35. Ban, L.; Scaloni, A.; Brandazza, A.; Angeli, S.; Zhang, L.; Pelosi, P. Chemosensory proteins of *Locusta migratoria*. *Insect Mol. Biol.* **2003**, *12*, 125–134. [CrossRef] [PubMed]
36. Lin, X.; Mao, Y.; Zhang, L. Binding properties of four antennae-expressed chemosensory proteins (CSPs) with insecticides indicates the adaptation of *Spodoptera litura* to environment. *Pest Biochem. Physiol.* **2018**, *146*, 43–51. [CrossRef]

37. Wanner, K.W.; Willis, L.G.; Theilmann, D.A.; Isman, M.B.; Feng, Q.; Plettner, E. Analysis of the insect os-d-like gene family. *J. Chem. Ecol.* **2004**, *30*, 889–911. [CrossRef]
38. Liu, G.X.; Arnaud, P.; Offmann, B.; Picimbon, J.F. Genotyping and bio-sensing chemosensory proteins in insects. *Sensors* **2017**, *17*, 1801. [CrossRef]
39. Honeybee Genome Sequencing Consortium. Insights into social insects from the genome of the honeybee *Apis mellifera*. *Nature* **2006**, *443*, 931–949. [CrossRef]
40. Forêt, S.; Wanner, K.W.; Maleszka, R. Chemosensory proteins in the honeybee: Insights from the annotated genome, comparative analysis and expression profiling. *Insect Biochem. Mol. Biol.* **2007**, *37*, 19–28. [CrossRef]
41. Wallberg, A.; Bunikis, I.; Vinnere Pettersson, O.; Mosbech, M.B.; Childers, A.K.; Evans, J.D.; Mikheyev, A.S.; Robertson, H.M.; Robinson, G.E.; Webster, M.T. A hybrid de novo genome assembly of the honeybee, *Apis mellifera*, with chromosome length scaffold. *BMC Genomics* **2019**, *20*, 275. [CrossRef]
42. Desjardins, C.A.; Oliveira, D.C.S.G.; Edwards, R.M.; Dang, P.M.; Lee, D.; Colbourne, J.K.; Tettelin, H.; Hunter, W.B.; Werren, J.H. Nasonia Expression Libraries from Larvae, Pupae, and Adults. 2007. Available online: NasoniaBase, http://hymenopteragenome.org.
43. Werren, J.H.; Richards, S.; Desjardins, C.A.; Niehuis, O.; Gadau, J.; Colbourne, J.; Beukeboom, L.W.; Desplan, C.; Elsik, C.G.; Grimmelikhuijzen, C.J.P.; et al. Functional and evolutionary insights from the genomes of three parasitoid *Nasonia* species. *Science* **2010**, *327*, 343–348. [CrossRef]
44. Bonasio, R.; Zhang, G.; Ye, C.; Mutti, N.; Fang, X.; Qin, N.; Donahue, G.; Yang, P.; Li, Q.; Li, C.; et al. Genomic comparison of the ants *Camponotus floridanus* and *Harpegnathos saltator*. *Science* **2010**, *329*, 1068–1071. [CrossRef] [PubMed]
45. Kulmuni, J.; Wurm, Y.; Pamilo, P. Comparative genomics and chemosensory protein genes reveals rapid evolution and positive selection in ant-specific duplicates. *Heredity* **2013**, *110*, 538–547. [CrossRef] [PubMed]
46. Sabatier, L.; Jouanguy, E.; Dostert, C.; Zachary, D.; Dimarcq, J.L.; Bulet, P.; Imler, J.L. Pherokine-2 and -3: Two *Drosophila* molecules related to pheromone/odor-binding proteins induced by viral and bacterial infections. *Eur. J. Biol.* **2003**, *270*, 3398–33407. [CrossRef] [PubMed]
47. Voordekers, K.; Verstrepen, K. Experimental evolution of the model eukaryote *Saccharomyces cerevisiae* yields insight into the molecular mechanisms underlying adaptation. *Curr. Opin. Microbiol.* **2015**, *28*, 1–9. [CrossRef]
48. Chen, T.; Reith, M.E.; Ross, N.W.; MacRae, T.H. Expressed sequence tag (EST)-based characterization of gene regulation in *Artemia* larvae. *Invert. Rep. Dev.* **2003**, *44*, 33–44. [CrossRef]
49. Gottfried, J.A.; Wilson, D.A. Smell. In *Neurobiology of Sensation and Reward*; Gottfried, J.A., Ed.; CRC Press/Taylor & Francis: Boca Raton, FL, USA, 2011.
50. Celniker, S.E.; Wheeler, D.A.; Kronmiller, B.; Carlson, J.W.; Halpern, A.; Patel, S.; Adams, M.; Champe, M.; Dugan, S.P.; Frise, E.; et al. Finishing a whole-genome shotgun: Release 3 of the *Drosophila melanogaster* euchromatic genome sequence. *Genome Biol.* **2002**, *3*. [CrossRef]
51. Stapleton, M.; Carlson, J.; Brokstein, P.; Yu, C.; Champe, M.; George, R.; Guarin, H.; Kronmiller, B.; Pacleb, J.; Park, S.; et al. A *Drosophila* full-length cDNA resource. *Genome Biol.* **2002**, *3*. [CrossRef]
52. Stapleton, M.; Liao, G.; Brokstein, P.; Hong, L.; Carninci, P.; Shiraki, T.; Hayashizaki, Y.; Champe, M.; Pacleb, J.; Wan, K.; et al. The *Drosophila* gene collection: Identification of putative full-length cDNAs for 70% of *D. melanogaster* genes. *Genome Res.* **2002**, *12*, 1294–1300. [CrossRef]
53. Mita, K.; Morimyo, M.; Okano, K.; Koike, Y.; Nohata, J.; Kawasaki, H.; Kadono-Okuda, K.; Yamamoto, K.; Suzuki, M.G.; Shimada, T.; et al. The construction of an EST database for *Bombyx mori* and its application. *Proc. Natl. Acad. Sci. USA* **2003**, *104*, 14121–14126. [CrossRef]
54. Ai, H. Sensors and sensory processing for airborne vibrations in silk moths and honeybees. *Sensors* **2013**, *13*, 9344–9363. [CrossRef]
55. Ellison, C.; Brun, Y.V. Mechanosensing: A regulation sensation. *Curr. Biol.* **2015**, *25*, R113–R115. [CrossRef] [PubMed]
56. Topham, M.K.; Prescott, S.M. *Handbook of Cell Signaling*, 2nd ed.; Elsevier Academic Press: Cambridge, MA, USA, 2009.
57. Lohr, J.B.; Kuhn-Velten, W.N. Protein phosphorylation changes ligand-binding efficiency of cytochrome P450c17 (CYP17) and accelerates its proteolytic degradation: Putative relevance for hormonal regulation of CYP17 activity. *Biochem. Physiol. Res. Commun.* **1997**, *231*, 403–408. [CrossRef] [PubMed]

58. Helling, S.; Huttermann, M.; Ramzan, R.; Kim, S.H.; Lee, I.; Muller, T.; Langerfeld, E.; Meyer, H.E.; Kadenbach, B.; Vogt, S.; et al. Multiple phosphorylations of cytochrome c oxidase and their functions. *Proteomics* **2012**, *12*, 950–959. [CrossRef] [PubMed]
59. Mansilla, M.C.; de Mendoza, D. The *Bacillus subtilis* desaturase: A model to understand phospholipid modification and temperature sensing. *Arch. Microbiol.* **2005**, *183*, 229–235. [CrossRef]
60. Tang, G.Q.; Novitzky, W.P.; Griffin, H.C.; Huber, S.C.; Dewey, R.E. Oleate desaturase enzymes of soybean: Evidence of regulation through differential stability and phosphorylation. *Plant J.* **2005**, *44*, 433–446. [CrossRef]
61. Ohnishi, A.; Hull, J.; Kaji, M.; Hashimoto, K.; Lee, J.M.; Tsuneizumi, K.; Suzuki, T.; Dohmae, N.; Matsumoto, S. Hormone signaling linked to silkmoth sex pheromone biosynthesis involves Ca^{2+}/Calmodulin-dependent protein kinase II-mediated phosphorylation of the insect PAT family protein *Bombyx mori* lipid storage droplet protein-1 (BmLsD1). *J. Biol. Chem.* **2011**, *286*, 24101–24112. [CrossRef]
62. Du, M.; Yin, X.; Zhang, S.; Zhu, B.; Song, Q.; An, S. Identification of lipases involved in PBAN stimulated pheromone production in *Bombyx mori* using the DGE and RNAi approaches. *PLoS ONE* **2012**, *7*, e31045. [CrossRef]
63. Liu, P.; Peng, H.J.; Zhu, J. Juvenile hormone-activated phospholipase C pathway enhances transcriptional activation by the methoprene-tolerant protein. *Proc. Natl. Acad. Sci. USA* **2015**, *112*, E1871–E1879. [CrossRef]
64. Enan, E.; Matsumura, F. Stimulation of protein phosphorylation in intact rat brain synaptosomes by a pyrethroid insecticide, deltamethrin. *Pest Biochem. Physiol.* **1991**, *39*, 182–195. [CrossRef]
65. Gollamudi, S.; Johri, A.; Callingasan, N.Y.; Yang, L.; Elemento, O.; Beal, M.F. Concordant signaling pathways produced by pesticide exposure in mice correspond to pathways identified in human Parkinson's disease. *PLoS ONE* **2012**, *7*, e36191. [CrossRef]
66. Yang, J.S.; Symington, S.; Clark, J.M.; Park, Y. Permethrin, a pyrethroid insecticide, regulates ERK1/2 activation through membrane depolarization-mediated pathway in HepG2 hepatocytes. *Food Chem. Toxicol.* **2018**, *121*, 387–395. [CrossRef]
67. Le Goff, G.; Giraudo, M. Effects of Pesticides on the Environment and Insecticide Resistance. In *Olfactory Concepts of Insect Control-Alternative to Insecticides*; Picimbon, J.F., Ed.; Springer Nature Switzerland: Basel, Switzerland, 2019; Volume 1, pp. 51–78.
68. Devillard, E.; McIntosh, F.M.; Duncan, S.H.; Wallace, R.J. Metabolism of linoleic acid by human gut bacteria: Different routes for biosynthesis of conjugated linoleic acid. *J. Bacteriol.* **2007**, *189*, 2566–2570. [CrossRef]
69. Takatsuka, J.; Nakai, M.; Shinoda, T. A virus carries a gene encoding juvenile hormone acid methyltransferase, a key regulatory enzyme in insect metamorphosis. *Sci. Rep.* **2017**, *7*. [CrossRef]
70. Nates, S.F.; McKenney, C.L., Jr. Growth, lipid class and fatty acid composition in juvenile mud crabs (*Rhithropanopeus harrisii*) following larval exposure to Fenoxycarb, insect juvenile hormone analog. *Comp. Biochem. Physiol. C Toxicol. Pharmacol.* **2000**, *127*, 317–325. [CrossRef]
71. Nagaraju, G.P.C. Reproductive regulators in decapod crustaceans: An overview. *J. Exp. Biol.* **2011**, *214*, 3–16. [CrossRef]
72. Tamone, S.L.; Harrison, J.F. Linking insects with crustacea: Physiology of the Pancrustacea: An introduction to the symposium. *Integr. Comp. Physiol.* **2015**, *55*, 765–770. [CrossRef]
73. Nijhout, H.F. *Insect Hormones*; Princeton University Press: Princeton, NJ, USA, 1994; p. 280.
74. Blomquist, G.J.; Dwyer, L.A.; Chu, A.J.; Ryan, R.O.; de Renobales, M. Biosynthesis of linoleic acid in a termite, cockroach and cricket. *Insect Biochem.* **1982**, *12*, 349–353. [CrossRef]
75. Aksenov, V.; Rollo, C.D. Necromone death cues and risk avoidance by the cricket *Acheta domesticus*: Effects of sex and duration of exposure. *J. Insect Behav.* **2017**, *30*, 259–272. [CrossRef]
76. Blaul, B.; Steinbauer, R.; Merkl, P.; Merkl, R.; Tschochner, H.T.; Ruther, J. Oleic acid is a precursor of linoleic acid and the male sex pheromone in *Nasonia vitripennis*. *Insect Biochem. Mol. Biol.* **2014**, *51*, 33–40. [CrossRef]
77. Rule, G.; Roelofs, W.L. Biosynthesis of sex pheromone components from linolenic acid in arctiid moths. *Arch. Insect Biochem. Physiol.* **1989**, *12*, 89–97. [CrossRef]
78. Sakai, R.; Fukuzawa, M.; Nakano, R.; Tatsuki, S.; Ishikawa, Y. Alternative suppression of transcription from two desaturase genes is the key for species-specific sex pheromone biosynthesis in two *Ostrinia* moths. *Insect Biochem. Mol. Biol.* **2009**, *39*, 62–67. [CrossRef]

79. Moto, K.; Suzuki, M.G.; Hull, J.J.; Kurata, R.; Takahashi, S.; Yamamoto, M.; Okano, K.; Imai, K.; Ando, T.; Matsumoto, S. Involvement of a bifunctional fatty-acyl desaturase in the biosynthesis of the silkmoth, *Bombyx mori*, sex pheromone. *Proc. Natl. Acad. Sci. USA* **2004**, *101*, 8631–8636. [CrossRef]
80. Vukašinović, E.L.; Pond, D.W.; Worland, M.R.; Kojić, D.; Purać, J.; Blagojević, D.P.; Grubor-Lajsić, G. Diapause induces changes in the composition and biophysical properties of lipids in larvae of the European corn borer, *Ostrinia nubilalis* (*Lepidoptera*: *Crambidae*). *Comp. Biochem. Physiol. B Biochem. Mol. Biol.* **2013**, *165*, 219–225. [CrossRef]
81. Briand, L.; Swasdipan, N.; Nespoulos, C.; Bézirard, V.; Blon, F.; Huet, J.C.; Ebert, P.; Pernollet, J.C. Characterization of a chemosensory protein (ASP3c) from honeybee (*Apis mellifera* L.) as a brood pheromone carrier. *Eur. J. Biochem.* **2002**, *269*, 4586–4596. [CrossRef]
82. Li, H.L.; Lou, B.G.; Cheng, J.A.; Gao, Q.K. The Chemosensory protein of Chinese honeybee, *Apis cerana cerana*: Molecular cloning of cDNA, immunocytochemical localization and expression. *Chin. Sci. Bull.* **2007**, *52*, 1355–1364. [CrossRef]
83. Pikielny, C.W.; Hasan, G.; Rouyer, F.; Rosbach, M. Members of a family of *Drosophila* putative odorant-binding proteins are expressed in different subsets of olfactory hairs. *Neuron* **1994**, *12*, 35–49. [CrossRef]
84. McKenna, M.P.; Hekmat-Scafe, D.S.; Gaines, P.; Carlson, J.R. Putative *Drosophila* pheromone-binding-proteins expressed in a subregion of the olfactory system. *J. Biol. Chem.* **1994**, *269*, 16340–16347.
85. Claridge-Chang, A.; Wijnen, H.; Naef, F.; Boothroyd, C.; Rajewsky, N.; Young, M.W. Circadian regulation of gene expression systems in the *Drosophila* head. *Neuron* **2001**, *32*, 657–671. [CrossRef]
86. McDonald, M.J.; Rosbach, M. Microarray analysis and organization of circadian gene expression in *Drosophila*. *Cell* **2001**, *107*, 567–578. [CrossRef]
87. Stathopoulos, A.; Van Drenth, M.; Erives, A.; Markstein, M.; Levine, M. Whole genome analysis of dorsal-ventral patterning in the *Drosophila* embryo. *Cell* **2002**, *111*, 687–701. [CrossRef]
88. Malcicka, M.; Visser, B.; Ellers, J. An evolutionary perspective on linoleic acid synthesis in animals. *Evol. Biol.* **2018**, *45*, 15–26. [CrossRef]
89. Guo, W.; Wang, X.; Ma, Z.; Xue, L.; Han, J.; Yu, D.; Kang, L. *CSP* and *Takeout* genes modulate the switch between attraction and repulsion during behavioral phase change in the migratory locust. *PLoS Genet.* **2011**, *7*, e1001291. [CrossRef]
90. Martín-Blázquez, R.; Chen, B.; Kang, L.; Bakkali, M. Evolution, expression and association of the chemosensory protein genes with the outbreak phase of two main pest locusts. *Sci. Rep.* **2017**, *7*, 6653. [CrossRef]
91. Whitfield, C.W.; Band, M.R.; Bonaldo, M.F.; Kumar, C.G.; Liu, L.; Pardinas, J.R.; Robertson, H.M.; Bento Soares, M.; Robinson, G.E. Annotated Expressed Sequence Tags and cDNA microarrays for studies of brain and behavior in the honey bee. *Genome Res.* **2002**, *12*, 555–566. [CrossRef]
92. Porath, H.T.; Hazan, E.; Shpigler, H.; Cohen, M.; Band, M.; Ben-Shahar, Y.; Levanon, E.Y.; Eisenberg, E.; Bloch, G. RNA editing is abundant and correlates with task performance in a social bumblebee. *Nat. Commun.* **2019**, *10*, 1605. [CrossRef]
93. Duan, Y.; Dou, S.; Huang, J.; Eisenberg, E.; Lu, J. Evolutionary forces on A-to-I RNA editing revealed by sequencing individual honeybee drones. *bioRXiv* **2020**. [CrossRef]
94. Alyokhin, A.; Baker, M.B.; Mota-Sanchez, D.; Dively, G.; Grafius, E. Colorado potato beetle resistance to insecticide. *Am. J. Potato Res.* **2008**, *85*, 395–413. [CrossRef]
95. Boyer, S.; Zhang, H.Y.; Lemperiere, G. A review of control methods and resistance mechanisms in stored-product insects. *Bull. Entomol. Res.* **2011**, *102*, 213–229. [CrossRef]
96. Mota-Sanchez, D.; Wise, J.C.; Vander Poppen, R.; Gut, L.J.; Hollingworth, R.M. Resistance of codling moth, *Cydia pomonella* (L.) (*Lepidoptera*: *Tortricidae*), larvae in Michigan to different insecticides with different modes of action and the impact on field residual activity. *Pest Manag. Sci.* **2010**, *64*, 881–890. [CrossRef]
97. Troczka, B.J.; Williamson, M.S.; Field, L.M.; Davies, T.G.E. Rapid selection for resistance to diamide insecticides in *Plutella xylostella* via specific amino acid polymorphisms in the ryanodine receptor. *Neurotoxicology* **2017**, *60*, 224–233. [CrossRef]
98. Luo, C.; Jones, C.M.; Devine, G.; Zhang, F.; Denholm, I.; Gorman, K. Insecticide resistance in *Bemisia tabaci* biotype Q (Hemiptera: Aleyrodidae) from China. *Crop Prot.* **2010**, *29*, 429–434. [CrossRef]
99. Liu, G.X.; Xuan, N.; Chu, D.; Xie, H.Y.; Fan, Z.X.; Bi, Y.P.; Picimbon, J.F.; Qin, Y.C.; Zhong, S.T.; Li, Y.F.; et al. Biotype expression and insecticide response of *Bemisia tabaci* chemosensory protein-1. *Arch. Insect Biochem. Physiol.* **2014**, *85*, 137–151. [CrossRef] [PubMed]

100. Taylor-Wells, J.; Senan, A.; Bermudez, I.; Jones, A.K. Species specific RNA A-to-I editing of mosquito RDL modulates GABA potency and influences agonistic, potentiating and antagonistic actions of ivermectin. *Insect Biochem. Mol. Biol.* **2018**, *93*, 1–11. [CrossRef]
101. Dong, K. Insect sodium channels and insecticide resistance. *Invert. Neurosci.* **2007**, *7*, 17–30. [CrossRef]
102. Chang, C.; Shen, W.K.; Wang, T.T.; Lin, Y.H.; Hsu, E.L.; Dai, S.M. A novel amino acid substitution in a voltage-gated sodium channel is associated with knockdown resistance to permethrin in *Aedes aegypti*. *Insect Biochem. Mol. Biol.* **2009**, *39*, 272–278. [CrossRef]
103. Claudianos, C.; Ranson, H.; Johnson, R.M.; Biswas, S.; Schuler, M.A.; Berenbaum, M.R.; Feyereisen, R.; Oakeshott, J.G. A deficit of detoxification enzymes: Pesticide sensitivity and environmental response in the honeybee. *Insect Mol. Biol.* **2006**, *15*, 615–636. [CrossRef]
104. Maleszka, J.; Forêt, S.; Saint, R.; Maleszka, R. RNAi-induced phenotypes suggest a novel role for a chemosensory protein CSP5 in the development of embryonic integument in the honeybee (*Apis mellifera*). *Dev. Genes Evol.* **2007**, *217*, 189–196. [CrossRef]
105. Zimmerlin, L.; Park, T.S.; Zambidis, E.T. Capturing human naïve pluripotency in the embryo and in the dish. *Stem Cells Dev.* **2017**, *26*, 1141–1161. [CrossRef]
106. Graf, T.; Stadtfeld, M. Heterogeneity of embryonic and adult stem cells. *Cell Stem Cell* **2008**, *3*, 480–483. [CrossRef]
107. Tweedel, K.S. The adaptability of somatic stem cells: A review. *J. Stem Cell Regen. Med.* **2017**, *13*, 3–13.
108. Pyza, E.M. Plasticity in invertebrate sensory systems. *Front. Physiol.* **2013**, *4*, 226. [CrossRef] [PubMed]
109. Fahrbach, S.E.; Van Nest, B.N. Synapsin-based approaches to brain plasticity in adult social insects. *Curr. Opin. Insect Sci.* **2016**, *18*, 27–34. [CrossRef] [PubMed]
110. Anton, S.; Gadenne, C. Effect of juvenile hormone on the central nervous processing of sex pheromone in an insect. *Proc. Natl. Acad. Sci. USA* **1999**, *96*, 5764–5767. [CrossRef]
111. Lechelt, J.L.; Evenden, M.L. Peripheral and behavioral plasticity of pheromone response and its hormonal control in a long-lived moth. *J. Exp. Biol.* **2009**, *212*, 2000–2006.
112. Barrozo, R.B.; Gadenne, C.; Anton, S. Switching attraction to inhibition: Mating-induced reversed role of sex pheromone in an insect. *J. Exp. Biol.* **2010**, *213*, 2933–2939. [CrossRef]
113. Barrozo, R.B.; Jarriault, D.; Deisig, N.; Gemeno, C.; Monsempes, C.; Lucas, P.; Gadenne, C.; Anton, S. Mating-induced differential coding of plant odour and sex pheromone in a male moth. *Eur. J. Neurosci.* **2011**, *33*, 1841–1850. [CrossRef]
114. Cheng, S.S.; Liu, J.Y.; Tsai, K.H.; Chen, W.J.; Chang, S.T. Chemical composition and mosquito larvicidal activity of essential oils from leaves of different *Cinnamomum osmophloeum* provenances. *J. Agric. Food Chem.* **2004**, *52*, 4395–4400. [CrossRef]
115. Baumann, K. Stem cells A key to totipotency. *Nat. Rev. Mol. Cell Biol.* **2017**, *18*, 137. [CrossRef]
116. Brooks, A.N.; Turkarslan, S.; Beer, K.D.; Lo, F.Y.; Baliga, N.S. Adaptation of cells to new environments. *Wiley Interdiscip. Rev. Syst. Biol. Med.* **2011**, *3*, 544–561. [CrossRef]
117. Picimbon, J.F. RNA/peptide editing in small soluble binding proteins: A new theory for the origin of life on earth's crust. *Preprints* **2020**, 2020010357.

Article

De Novo Transcriptome Identifies Olfactory Genes in *Diachasmimorpha longicaudata* (Ashmead)

Liangde Tang [1,†], Jimin Liu [2,†], Lihui Liu [2], Yonghao Yu [2], Haiyan Zhao [3,4,*] and Wen Lu [5,*]

1 Key Laboratory of Integrated Pest Management on Tropical Crops, Ministry of Agriculture and Rural Affairs, Environment and Plant Protection Institute, Chinese Academy of Tropical Agricultural Sciences, Haikou 571101, China; tangldcatas@163.com
2 Guangxi Key Laboratory for Biology of Crop Diseases and Insect Pests, Institute of Plant Protection, Guangxi Academy of Agricultural Sciences, Nanning 530007, China; ljimin@126.com (J.L.); yangmeiliu@yeah.net (L.L.); yxp1127@163.com (Y.Y.)
3 Department of Entomology, College of Tobacco Science, Guizhou University, Guiyang 550025, China
4 Guangxi Academy of Agricultural Sciences, Nanning 530007, China
5 College of Agriculture, Guangxi University, Nanning 530007, China
* Correspondence: haitianyiyan7611@163.com (H.Z.); luwenlwen@163.com (W.L.)
† Authors contribute equally.

Received: 3 January 2020; Accepted: 22 January 2020; Published: 29 January 2020

Abstract: *Diachasmimoorpha longicaudata* (Ashmead, *D. longicaudata*) (Hymenoptera: Braconidae) is a solitary species of parasitoid wasp and widely used in integrated pest management (IPM) programs as a biological control agent in order to suppress tephritid fruit flies of economic importance. Although many studies have investigated the behaviors in the detection of their hosts, little is known of the molecular information of their chemosensory system. We assembled the first transcriptome of *D. longgicaudata* using transcriptome sequencing and identified 162,621 unigenes for the Ashmead insects in response to fruit flies fed with different fruits (guava, mango, and carambola). We annotated these transcripts on both the gene and protein levels by aligning them to databases (e.g., NR, NT, KEGG, GO, PFAM, UniProt/SwissProt) and prediction software (e.g., SignalP, RNAMMER, TMHMM Sever). CPC2 and MIREAP were used to predict the potential noncoding RNAs and microRNAs, respectively. Based on these annotations, we found 43, 69, 60, 689, 26 and 14 transcripts encoding odorant-binding protein (OBP), chemosensory proteins (CSPs), gustatory receptor (GR), odorant receptor (OR), odorant ionotropic receptor (IR), and sensory neuron membrane protein (SNMP), respectively. Sequence analysis identified the conserved six Cys in OBP sequences and phylogenetic analysis further supported the identification of OBPs and CSPs. Furthermore, 9 OBPs, 13 CSPs, 3 GRs, 4IRs, 25 ORs, and 4 SNMPs were differentially expressed in the insects in response to fruit flies with different scents. These results support that the olfactory genes of the parasitoid wasps were specifically expressed in response to their hosts with different scents. Our findings improve our understanding of the behaviors of insects in the detection of their hosts on the molecular level. More importantly, it provides a valuable resource for *D. longicaudata* research and will benefit the IPM programs and other researchers in this filed.

Keywords: *Diachasmimorpha longicaudata*; Ashmead; parasitoid wasps; transcriptome; olfactory protein; odorant-binding protein; chemosensory protein

1. Introduction

Diachasmimoorpha longicaudata (Ashmead, *D. longicaudata*) is a solitary species of parasitoid wasp of several fruit fly species and has been introduced to many countries as a biological control agent. Its host, *Bactrocera dorsalis* Hendel, can attack many fruit species and some other plants, such as Caricaceae, Moraceae, Myrtaceae, Rosaceae, and Solaneaceae [1]. It is said that the female Ashmead insects can detect the fly larvae by sound in rotting fruit and that the attractant could be the fungal fermentation products rather than the chemical substances produced by the fly larvae [2,3]. Carrasco and colleagues reported that the presence of fly larvae was essential for the orientation of wasps [4]. Further, it was proposed that cues from the fruit can be used by the wasps directly and that the presence of the host can enhance the attraction towards a patch [5]. Interestingly, chemical compounds produced by the larvae can be detected by wasps to locate the host [6,7]. Once the female parasitoid is on the fruit, a specific chemical compound released by some Tephritidae species can be used to enhance the host search [7]. However, much is unknown about the chemosensory system of parasitoid wasps in response to their hosts.

Animals can use the chemosensory system to detect and discriminate chemical cues in the environment [8]. The chemical sensors of insects are mainly from the antennae system, which is a highly specific and extremely sensitive chemical detector, and olfactory proteins in the antennae can be used by the insects to detect very low-abundance odorants from thousands of odors and further to guide their behaviors, such as forage, mate hunting, host plant location, shelter, and selection of spawning sites [9]. Since the first odorant-binding protein (OBP) was identified in *Antheraea polyphemus* [10], research on insect OBP has become a hot spot in the field of entomology. However, OBP-related studies have mainly been demonstrated in some important pests [11], and very few have been reported in parasitoid wasps, which are very important natural enemies of the pests.

The chemoreception of insects involves three important events: (i) The uptake of signal molecules from the external environment; (ii) transport (diffusion) through the sensory hair; and (iii) interaction with the chemoreceptor, which in turn activates the cascade of events leading to spike activity in sensory neurons [12]. Some important protein families have been reported to participate in these events, such as OBP, sensory neuron membrane protein (SNMP), chemosensory protein (CSP), odorant receptor (OR), gustatory receptor (GR), and odorant ionotropic receptor (IR) [12]. Insect OBPs are small globular proteins (~135 to 220 amino acids) and are characterized by a specific domain that comprises six α-helices joined by three disulphide bonds [12]. They can be categorized into two subgroups: Pheromone-binding proteins, which are mainly distributed in the male antenna, and general OBP (GOBP), which can be found in multiple tissues of male and female insects and function in the recognition of odorants from plants and other animals [13]. There are about 300 OBPs in the NCBI database and not many studies have been demonstrated to identify the OBPs in parasitoid wasps. Xu and colleagues used transcriptome sequencing and identified 1 CSP, 21 OBPs, 53 ORs, 29 IRs, and 4 SNMPs in *Bactrocers minax* [14], an oligophagous tephritid insect whose host selection, and oviposition behavior largely depend on the perception of chemical cues. Zhu et al. reported Sgua-OBP1 and Sgua-OBP2 in *Scleroderma guani* (Hymenoptera: Bethylidae) and NvitOBP in *Nasonia vitripennis* [15]. Zhang identified 10 OBPs in *Microplitis mediator* (Halidag) [16]. Zhao et al. identified 25 OBPs, 80 ORs, 10 IRs, 11 CSP, 1 SNMPs, and 17 GRs in adult male and female *Chouioia cunea* antennae [17]. However, little is known about the olfactory proteins in *D. longicaudata*.

Harbi and colleagues demonstrated a multistep assay (e.g., olfactory, laboratory, and semi-field trials) and reported the preference of medfly-infected fruits, including apple, orange, peach, and clementine mandarins [18]. This experiment supports that different olfactory genes are expressed in response to different fruit scents. In this study, we used transcriptome sequencing to study the olfactory genes in *D. longicaudata*. By similarity, we identified a number of OBPs, CSPs, ORs, IRs, SNMPs, and GRs expressed in the Ashmead insects. Our results also showed that different olfactory genes were expressed in the search of their host with different fruit scents. This is the first time to identify the gene and protein sequences for the olfactory products in this species. Our findings will provide a basis for future molecular studies and improve our understanding of the chemosensory system of parasitoid wasps.

2. Materials and Methods

2.1. Insect Rearing

The Ashmead and fruit fly larvae were obtained from the Institute of Plant Protection (IPP), Hainan Academy of Agricultural Sciences and maintained in the experimental fields. The fruit flies were fed with guava (G), mango (M), and carambola (C) separately. A mixture of yeast and sucrose (1:1) was used as supplementary nutrition for the fruit flies in the adult stage. Then, late-second and early-third instar fruit flies were used as hosts for the Ashmead insects. The Ashmead insects were fed with 15% honey water and clean water; the fifth, sixth and seventh generations of the Ashmead adults, which were parasitic to the fruit flies, maintained with G (G1~G3), M (M1~M3), and C (C1~C3), were used as biological replicates. The antenna, head, breast, abdomen, and feet tissues of the one male and two female wasps were mixed together for RNA extraction.

2.2. RNA Isolation, Library Construction, and Deep Sequencing

Total RNA was extracted from the insect tissues using the TRIzol reagent, as previously described [19,20]. The quality and quantity of total RNA were determined by multiple instruments, including a Nanodrop 2000 (Thermo Scientific, MA, USA), Qubit 4 Fluorometer (Invitrogen, CA, USA), and Agilent 2100 Bioanalyzer. Then, the total RNA (1 μg) of each sample was used to build the cDNA library using the TruSeq RNA Library Preparation Kit v2 protocol (Illumina, CA, USA), as described [20]. After the cDNA libraries were quality controlled by the Agilent 2100 Bioanalyzer and qRT-PCR, they were sequenced on the Illumina HiSeqXTEN platform with the paired-end 150 strategy. Raw sequencing reads of these samples can be accessed from the NCBI SRA platform under the accession numbers SRR10766480~SRR10766488.

2.3. Transcriptome Assembly

Raw sequencing reads were cleaned using the trim_galore v0.5.0 and the clean data was quality controlled by FASTQC v0.11.7 (http://www.bioinformatics.babraham.ac.uk/projects/fastqc/). Then, clean reads were used to assemble the transcriptome for each sample using Trinity software with default parameters, as described [21]. In detail, high quality RNA-Seq reads were used to generate overlapping k-mers (25) and Inchworm was used to assemble sorted k-mers into transcript contigs based on the (k-1)-mer overlaps. Next, Chrysalis was used to cluster related Inchworm contigs into components by using grouped raw reads and paired read links. Then, a de Bruijn graph for each cluster was built by Chrysalis and reads were partitioned among the clusters. Finally, Butterfly was used to process the individual graphs and ultimately report the full-length transcripts. To remove redundant sequences, CD-HIT was used to cluster the assembled highly similar transcripts into Unigenes [22], which can be accessed in the NCBI TAS platform under the accession number GIF00000000. BUSCO v4 was used to evaluate the completeness of the assembled Unigenes [23] using the eukaryota_odb10 dataset.

2.4. Annotation for the Transcriptome

We annotated the assembled transcriptome by aligning them to different databases. Initially, the transcriptome was searched against the NR (Non Redundant Protein Sequence Database), NT (Nucleotide Sequence Database), KEGG (Kyoto Encyclopedia of Genes and Genomes) pathway, gene ontology (GO), Pfam, and UniProt/SwissProt databases using BLAST software, and hits with an e-value $> 1 \times 10^{-5}$ were filtered. Then, BLAST2GO was used to retrieve the GO annotation in terms of the biological process, cellular component, and molecular function [24]. Using the enzyme commission numbers produced by BLAST2GO, we mapped the assembled transcriptome to the KEGG pathway database and obtained the pathway annotation. rRNA transcripts were predicted using RNAMMER [25].

2.5. Likely Protein Identification and Annotation

We next extracted the likely proteins from the assembled transcriptome using TransDecoder. Then, the likely proteins were searched against the UniProtKB/Swiss-Prot database to identify known proteins, functional PFAM domains were identified using HMMER [26], signal peptides were predicted using SignalP [27], and transmembrane domains were predicted using TMHMM Sever v2.0 [28]. The EggNOG database v4.1 [29] was searched against to identify proteins in EuKaryotic Orthologous Groups (KOG), Clusters of Orthologous Groups (COGs), and non-supervised orthologous groups (NOGs). All the annotation for the assembled genes and likely proteins were subjected to the Trinotate v3.1.1 (http://trinotate.github.io) for combination.

Based on the gene annotation and likely protein annotation, we obtained the Unigenes, which were annotated into olfactory gene families, such as OBP, OR, IR, GR, and SNMP, using their names as key words. For the CSP transcripts, we aligned the Unigenes to all the CSP transcripts from NCBI GenBank. Hits with an e-value > 1×10^{-5} were filtered.

2.6. Noncoding Genes and microRNA Genes

Unannotated Ashmead genes were processed by the Coding Potential Calculator (CPC v2) with default parameters to identify potential long noncoding genes [30]. To identify potential microRNA (miRNA) genes, we first mapped all the animal mature miRNAs to the noncoding genes with a maximal of two mismatches [31]. Then, MIREAP was used to predict the miRNA precursors and MIRANDA was used to predict the target genes of these miRNAs [32].

2.7. Differential Expression Analysis

We aligned the clean reads of each sample to the Unigenes using Bowtie2 and profiled the gene expression using RSEM [33]. The trimmed mean of the M-values (TMM) method was used for normalization and edgeR was used to identify differentially expressed genes with the following cut-offs [34]: Count > 5, $\log_2$ fold change ($\log_2$ fc) >1 or $\log_2$ fc < −1, p-value < 0.05, and false discovery rate (FDR) < 0.05.

2.8. Function Enrichment Analysis

We calculated the p-value (calculated by Fisher's exact test) and q-value (calculated by the R package 'qvalue') for each GO term and KEGG pathway involved in the differentially expressed genes. Enriched terms should satisfy the following criteria: p-value < 0.05 and q-value < 0.05.

2.9. Phylogenetic Analysis

Phylogenetic trees were reconstructed for OBPs and CSPs using MEGA7 software [35]. We obtained the likely protein sequences for the top 5 highly expressed OBP and CSP transcripts. These sequences together with the homology protein sequences, obtained from NCBI, were subjected to MEGA7 to create phylogenetic trees using the neighbor-joining method. The bootstrap procedure based on 1000 replicates was used to assess node support, and the node support values < 50% were not shown. Figtree v1.4.3 (https://github.com/rambaut/figtree/) was used to visualize the results.

2.10. qRT-PCR Verification

We used real-time quantitative reverse transcription polymerase chain reaction (qRT-PCR) to validate the expression levels of three randomly selected transcripts (TRINITY_DN1020_c0_g1_i3, TRINITY_DN1284_c0_g1_i11, and TRINITY_DN500_c0_g1_i2). Forward and reverse primers of the three transcripts and the internal control (β-actin) were predicted using Prime3 and synthesized at BGI-Shenzhen. The procedure of the qRT-PCR experiment was the same as the previous study [21]. Each transcript was measured three times in every sample and three independent repeats were performed (n = 9). The Delta cycle threshold (ΔCt) was used to present the expression of a transcript in the sample and ΔΔCt was used to show the expression difference between two samples. We used the relative normalized expression (RNE) to show the expression changes: $RNE = 2^{-Ct}$.

3. Results

3.1. Animal and Transcriptome Sequencing

After the Ashmead animals (three females and one male) were maintained with oriental fruit flies, which were fed guava (G), mango (M), and carambola ©, the antenna, head, breast, abdomen, and feet tissues were mixed together for RNA extraction and transcriptome sequencing. After data cleaning, we obtained a total of ~622.32 million reads (~69.15 million reads on average) and assembled 24,201 to 34,302 genes using Trinity for all the samples (Table 1). After similar genes/transcripts were clustered and merged, we finally obtained 162,621 Unigenes for the Ashmead transcriptome, with an average length of 1425.14 bp. The N50, GC content, and size of the transcriptome were calculated as 3572, 41.88%, and ~231 M, respectively (Table 1). Length distribution analysis showed that 53.05% of the total transcripts were longer than 500 bp and 11,101 transcripts (6.83% of the total transcripts) were longer than 5000 bp (Figure 1A). Last, we used BUSCO to evaluate the completeness of the assembled Unigenes and the results showed 99.6% of the assembled Unigenes were complete. In detail, out of the 255 evaluated BUSCOs in the dataset, 254 were complete, including 228 duplicated and 26 single-copy BUSCOs.

Table 1. Overview of the de novo transcriptome of Ashmead.

	G1	G2	G3	M1	M2	M3	C1	C2	C3
Clean reads	67,492,576	74,983,476	70,212,518	78,796,858	61,848,876	72,961,796	71,005,592	54,703,286	70,316,132
Assembled genes	34,302	26,032	25,398	24,201	33,375	33,081	33,308	24,772	24,707
Assembled transcripts	49,914	40,912	39,528	38,089	50,517	51,704	50,530	38,856	39,637
Unigenes					162,621				
Mean Length (bp)					1425.14				
N50					3572				
GC (%)					41.88				
Total bases					231,757,796				
Expressed transcripts	10,736	10,320	10,645	9860	10,863	11,695	16,827	11,765	14,728

3.2. Transcriptome Annotation

We first annotated the assembled Ashmead transcriptome on the transcript level. All the transcripts were aligned to public databases for full annotation and Figure 1B showed 74,264, 68,185, 54,536, 58,109, 61,794, 58,113, and 22,546 transcripts were aligned to the NR, NT, UniProt/SwissProt, KEGG pathway, KOG, Pfam, and GO databases, respectively. Tthee NR mapping results (Figure 1C) showed the top 10 species aligned by the assembled Ashmead transcripts and the majority of the transcripts were aligned to *Diachasma alloeum* (45,569 transcripts), *Fopius arisanus* (5217 transcripts), and *Rhinolophus sinicus* (4920 transcripts). Unsurprisingly, the top two species together with Ashmead were all from the Braconidae family. GO annotation revealed that 11,131, 7722, and 9078 transcripts were involved in "binding", "membrane", and "cellular process", respectively (Figure 1C). Then, we categorized the KEGG pathway annotation (Figure 1E) into six groups: Cellular progresses, environmental information processing, genetic information processing, human diseases, metabolism, and organism systems. Among them, "signal transduction" is the most significant pathway, which involved 9231 transcripts. KOG annotation also revealed that 14,221 transcripts were involved in the signal transduction mechanisms (Figure 1F).

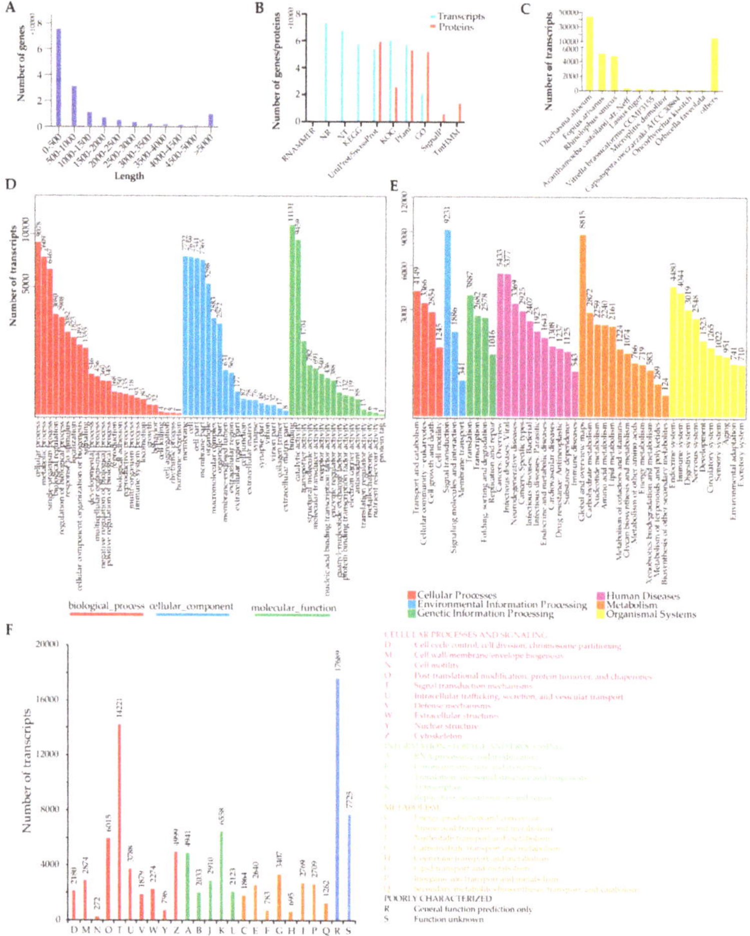

Figure 1. Overview of the assembled Ashmead transcriptome and annotation. (**A**) Length distribution of the assembled Ashmead transcriptome. (**B**) Number of transcripts and likely proteins aligned to different databases. (**C**) Number of the transcripts aligned to other species in the NR mapping results. (**D**) GO annotation for the assembled Ashmead transcriptome. (**E**) KEGG pathway annotation for the assembled Ashmead transcriptome. (**F**) KOG annotation for the assembled transcriptome.

Next, TranDecoder predicted 88,215 likely proteins encoded by the assembled Ashmead transcriptome and we annotated them using the Trinotate pipeline. It was shown that 59,887, 26,402, 53,725, and 52,572 were aligned to the UniProt/SwissProt, KOG, Pfam, and GO databases, respectively (Figure 1B). In addition, 6642 and 14,672 likely proteins were predicted to contain signal peptides and transmembrane helices (Figure 1B).

In addition, we predicted 74,963 of the unannotated transcripts using CPC2 had low coding probability. Interestingly, 2128 of the unannotated transcripts were predicted to be possible coding transcripts, which might be specific to the Ashmead and require further experiments to be verified. Next, we predicted 87 transcripts have the potential of producing miRNAs (Table S1). Notably, 84 of them were probably derived from the intron regions of coding genes while 1 and 3 derived from the Ashmead specific coding gene and long noncoding genes, respectively.

3.3. Olfactory Genes

We next identified genes encoding the olfactory gene from five families, including OBP, CSP, OR, GR, IR, and SNMP. In the Ashmead transcriptome, we found 43 transcripts encoding OBPs (Table 2, Table S2) by similarity and 35 of them were predicted to have complete ORFs by TransDecoder. The Ashmead OBPs were categorized into four sub-families: OBP-56, -69, -72, and -83 (Table S2). Multiple sequence alignment (Figure S1) showed these OBPs have 6 conserved cysteine residues (Cys) and SignalP predicted that 26 of these OBPs have the signal peptides located in the first 23 amino acids (aa) and 2 located in the first 37 aa. We identified that 69 transcripts had the potential of encoding CSPs in the Ashmead transcriptome by aligning the likely proteins to the known CSPs (Table 2, Table S2) and 63 of them had intact ORFs. SignalP identified 58 CSPs with the signal peptides in the first 28 aa while transmembrane domains were found in 13 CSPs. Surprisingly, 689 Ashmead transcripts had the capacity of encoding ORs and 115 transcripts were found to encode OR-13 (Table 2, Table S2). We also identified nine transcripts encoding OR coreceptors (OR-co). Out of the 347 OR transcripts that had intact ORFs, 276 were predicted to have transmembrane domains (Table 2). In the Ashmead transcriptome, there were 26 transcripts encoding IRs, including 3 IR-21, 18 IR-25, 4 IR-68, and 2 IR-93 (Table 2). Two thirds of the IR transcripts that had intact ORFs were predicted to have transmembrane domains. In addition, we identified 60 GR and 14 SNMP transcripts (Table 2). GR-28 and SNMP-1 were the largest group, which corresponded 19 and 11 transcripts, respectively. Further, we found 55 transcripts that had either the 7tm chemosensory receptor (PF02949) or GOBP (PF01395) PFAM domain in their protein sequences (Table 2, Table S2).

Table 2. Olfactory genes identified in the Ashmead transcriptome.

Type	Transcripts	Sub-Family	Intact ORF	SignalP	TMHMM
OBP	43	21 OBP-56, 7 OBP-69, 4 OBP-72, 11 OBP-83	35	28	15
CSP	69	5 CSP-1, 3 CSP-3, 2 CSP-4, 5 CSP-5, 10 CSP-6, 3 CSP-7, 4 CSP-8, 37 CSP *	63	50	13
OR	689	45 OR-1, 6 OR-1F12, 21 OR-10, 115 OR-13, 42 OR-2, 46 OR-22, 2 OR-23, 9 OR-24, 2 OR-245, 1 OR-260, 2 OR-266, 1 OR-277, 15 OR-30, 6 OR-33, 31 OR-4, 5 OR-42, 14 OR-43, 2 OR-45, 24 OR-46, 47 OR-47, 26 OR-49, 2 OR-59, 1 OR-5, 1 OR-63, 48 OR-67, 1 OR-69, 6 OR-71, 13 OR-7, 1 OR-81, 62 OR-82, 37 OR-85, 11 OR-92, 11 OR-94, 5 OR-98, 4 OR-9, 9 OR-co, 1 OR-142, 14 OR *	347	25	276
IR	26	3 IR-21, 18 IR-25, 4 IR-68, 2 IR-93	18	4	12
GR	60	6 GR-2, 5 GR-107, 3 GR-15, 1 GR-23, 19 GR-28, 4 GR-2, 8 GR-43, 2 GR-64, 3 GR-66, 8 GR *	26	1	25
SNMP	14	11 SNMP-1, 3 SNMP-3	10	1	9
Others [a]	54	16 CR, 38 GOBP	10	12	22

[a] Transcripts containing the chemosensory receptor or the GOBP domain, by PFAM annotation. * Products are not specified.

3.4. Gene Expression Profile

We next profiled the gene expression in the Ashmead insects maintained with the fruit flies fed with the three kinds of fruits. After lowly expressed genes (count < 5) were filtered, RSEM identified a total of 63,627 transcripts in the Ashmead animals, of which 46,607, 49,253, and 53,558 transcripts were distributed in G, M, and C, respectively (Table S3). The Venn diagram (Figure 2A) revealed 38,009 transcripts commonly expressed in all samples while 3115, 4571, and 8159 were specifically detected in G, M, and C, respectively. Interestingly, not all the olfactory genes were expressed in the insects and we found 39 OBPs, 65 CSPs, 29 GRs, 382 ORs, 15 IRs, and 13 SNMPs Figure 2B,C showed the expression levels of these olfactory transcripts in the insects, and revealed that different olfactory genes of Ashmead insects are responsible for the fruit flies with different fruits. According to the average expression levels, we showed the top five highly expressed olfactory transcripts identified in this study (Table 3) and it was revealed that the identities of highly expressed olfactory transcripts were shared by the parasitoid wasps of the fruit flies fed with different fruits. Notably, OBP56 and OBP69 were highly expressed in the insects; IR25a and SNMP1 were the only highly expressed transcript for the IR and SNMP groups, respectively.

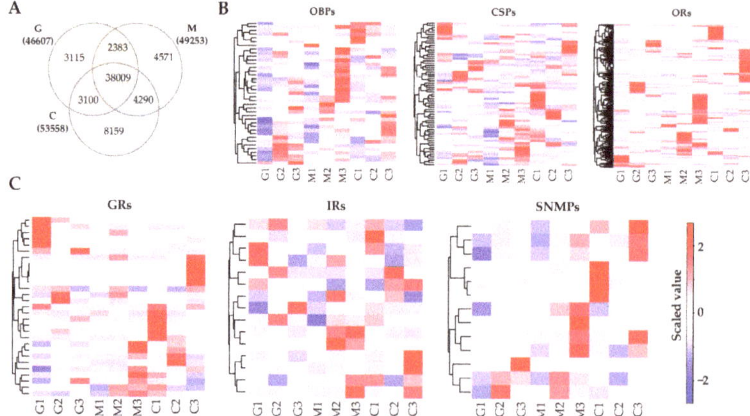

Figure 2. Expression levels of olfactory transcripts in the Ashmead insects. (**A**) Number of transcripts identified in the insects stimulated by three fruits. (**B**) Heat maps of the expression levels of OBP, CSP, and OR transcripts. (**C**) Heat maps showing the expression levels of transcripts encoding GRs, IRs, and SNMPs in the Ashmead insects.

3.5. Phylogenetic Analysis

We next compared the sequences of olfactory proteins identified in Ashmead and some other homology organisms and analyzed their phylogenetic relationship. First, we constructed the phylogenetic tree for the top 5 OBPs (Table 3) and 28 other OBP sequences obtained from NCBI. Detailed information, including accession numbers and species, can be accessed in Table S3. The phylogenetic tree of OBPs (Figure 3A) showed high similarity (68.9% to 96.5%) between Ashmead OBPs and other species, such as *Aethina tumida* (Atumi), *Aphidius gifuensis* (Agifu), *Cephus cinctus* (Ccinc), *Cotesia chilonis* (Cchil). *Diachasma alloeum* (Dallo), *Fopius arisanus* (Faris), *Megachile rotundata* (Mrotu), *Meteorus pulchricornis* (Mpulc), *Microplitis demolitor* (Mdemo), and *Microplitis mediator* (Mmedi). We next performed the phylogenetic analysis for CSPs. The top 5 highly expressed CSPs (Table 3) were compared with 22 CSPs from other species, like *Vespa velutina* (Vvelu), Ccinc, *Sclerodermus*, Mpulc, and *Yemma signatus* (Ysign). Detailed accession numbers of these CSPs can be accessed in Table S3. It is clear that the Ashmead CSPs can be clustered with other known CSPs. The phylogenetic analysis supported the identification and characterization of OBP and CSP transcripts in this study.

3.6. Differential Expression Analysis

Another important goal of this study was to identify genes in the insects in response to their parasitic hosts, which had difference fruit scents. Using edgeR, we identified a total of 2650 transcripts differentially expressed in the Ashmead insects in response to the fruit flies with different scents (Table S4), and the number of differentially expressed transcripts can be seen in Figure 4A. Compared to G, there were 1466 upregulated and 53 downregulated transcripts identified in both C and M (Figure 4B). Some transcripts were specifically expressed in the insects when they were parasitic to the fruit flies with one fruit scent (Figure 4B). We next analyzed the differentially expressed transcripts encoding olfactory proteins in the Ashmead insects in response to the three fruit flies. A total of 58 transcripts encoding olfactory proteins were found, including 9 OBPs, 13 CSPs, 3 GRs, 4IRs, 25 ORs, and 4 SNMPs (Figure 4C). This evidence further supports the existence of multiple pathways of Ashmead insects in response to different fruits scents. In addition to olfactory proteins, some other protein families were differentially expressed in the wasps maintained with fruit flies supplied with different fruits, such as 2 LOC107047718 (Putative 7 transmembrane sweet-taste receptor of 3 gcpr), 88 ribosomal proteins, 109 transcription factors, 99 histones, 11 heat shock proteins, and 43 G-protein coupled receptors/regulators (Table S4). The differential expression of transcripts from different families indicated the complicated regulation mechanisms of parasitoid wasps in response to their hosts with different fruit scents. More experiments are required to explore their functions in this process.

Table 3. Top five transcripts encoding olfactory proteins in Ashmead insects.

TranscriptID	G [a]	M [a]	C [a]	Protein/Gene	Description
OBP					
TRINITY_DN22_c0_g3_i1	41,518.89	55,018.99	39,228.41	OB56D	General odorant-binding protein 56d
cluster_contig4670	20,459.11	27,957.59	83,228.40	OB56H	General odorant-binding protein 56h
cluster_contig3738	15,552.04	13,752.58	20,997.13	OB69A	General odorant-binding protein 69a
TRINITY_DN3641_c0_g2_i1	5,010.50	13,074.19	4,838.31	OB69A	General odorant-binding protein 69a
TRINITY_DN23528_c0_g1_i1	3853.93	5367.95	6228.73	OB56D	General odorant-binding protein 56d
CSP					
TRINITY_DN1018_c0_g1_i5	11,065.74	15,620.31	10,030.62	CSP	chemosensory protein
TRINITY_DN661_c1_g1_i1	9447.67	10,194.33	7724.67	THK33221.1	chemosensory protein 4
TRINITY_DN4258_c0_g1_i1	4538.67	4738.33	5976.67	AZQ24964.1	chemosensory protein, partial
TRINITY_DN1848_c0_g1_i6	5121.01	3247.27	4919.50	CSP	chemosensory protein
cluster_contig14748	3679.27	3944.22	2395.00	THK33222.1	chemosensory protein 5
OR					
cluster_contig6330	41,364.85	68,243.05	24,104.78	LOC107043577	Odorant receptor
TRINITY_DN1357_c0_g1_i5	18,042.97	23,565.93	11,831.46	OR43A	Odorant receptor 43a
TRINITY_DN1357_c0_g1_i6	7765.79	27,596.19	13,421.25	OR43A	Odorant receptor 43a
cluster_contig12468	16,698.80	23,528.66	8172.54	LOC107043576	Odorant receptor
cluster_contig9759	2835.24	2248.79	3479.03	OR43A	Odorant receptor 43a
GR					
TRINITY_DN9119_c0_g1_i2	455.68	489.30	545.92	GR107	gustatory receptor Gr107
TRINITY_DN3525_c0_g1_i10	288.94	333.98	230.17	GR43A	gustatory receptor for sugar taste 43a-like
cluster_contig36	290.10	43.82	30.95	GR43a	gustatory receptor for sugar taste 43a-like
TRINITY_DN14061_c0_g1_i3	91.79	116.05	78.74	GR43a	gustatory receptor for sugar taste 43a-like
TRINITY_DN4338_c0_g1_i4	90.22	77.22	64.01	GR107	gustatory receptor Gr107
IR					
TRINITY_DN11328_c0_g1_i1	41.09	39.27	45.12	IR25A	Ionotropic receptor 25a
cluster_contig8505	39.77	34.65	39.58	IR25A	Ionotropic receptor 25a
TRINITY_DN4424_c0_g1_i2	16.26	12.92	47.84	IR25A	Ionotropic receptor 25a
cluster_contig11752	19.21	25.57	18.80	IR25A	Ionotropic receptor 25a
TRINITY_DN13912_c1_g2_i1	20.88	10.71	20.37	IR25A	Ionotropic receptor 25a
SNMP					
TRINITY_DN1934_c0_g1_i1	3567.61	3632.31	5117.35	SNMP1	Sensory neuron membrane protein 1
TRINITY_DN4669_c1_g1_i1	176.96	204.92	206.30	SNMP1	Sensory neuron membrane protein 1
TRINITY_DN1867_c0_g2_i1	86.72	83.98	50.41	XP_015114627.1	Sensory neuron membrane protein 1-like
cluster_contig4264	70.20	73.23	46.97	SNMP1	Sensory neuron membrane protein 1
cluster_contig237	44.54	90.10	34.07	SNMP1	Sensory neuron membrane protein 1

[a] Normalized expression, TMM.

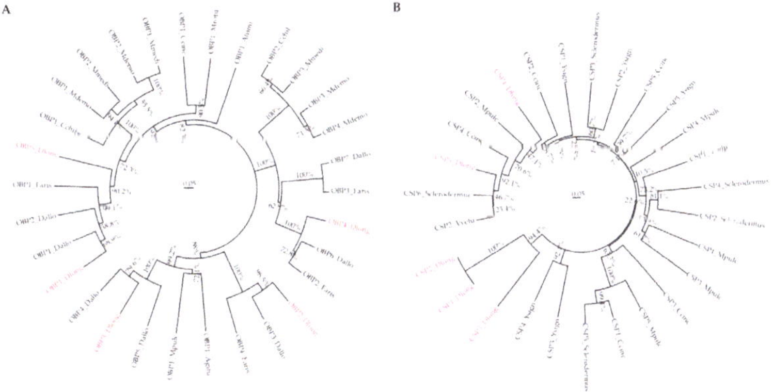

Figure 3. Phylogenetic trees for OBPs (**A**) and CSPs (**B**). Ashmead proteins are highlighted in red, the percentage represents the bootstrap value, and the scale bar represents the evolutionary distance.

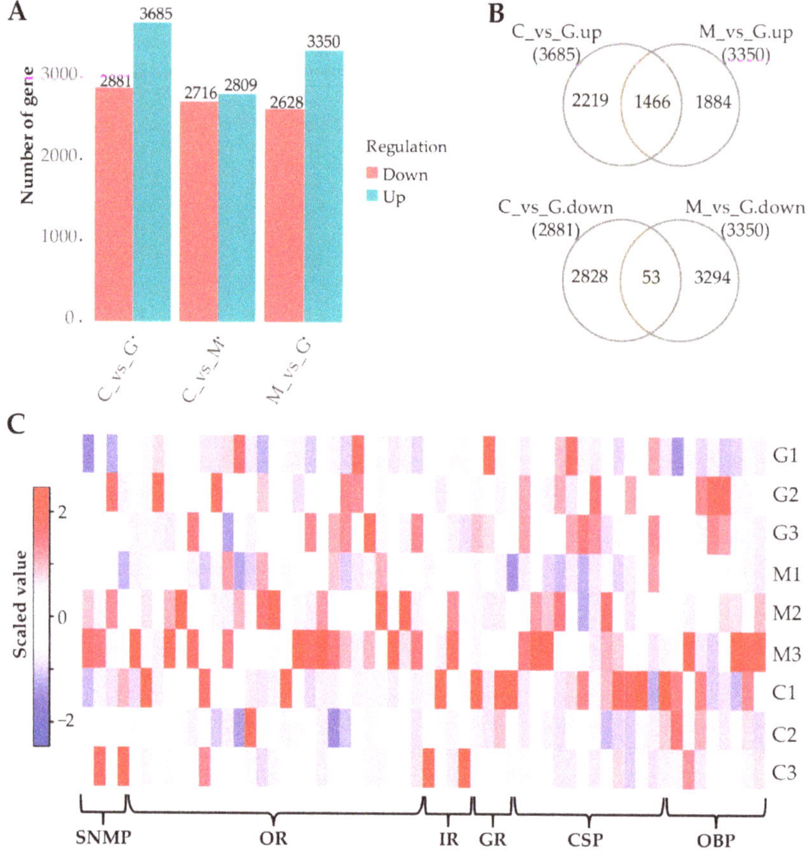

Figure 4. Differentially expressed transcripts in the parasitoid wasps of fruit flies fed with different fruits. (**A**) Number of differentially expressed transcripts in the insects in response to two fruit scents. (**B**) Venn diagram of up- (upper panel) and downregulated (lower panel) transcripts identified in C and M, compared to G. (**C**) A heat map showed the differential expression of 9 OBPs, 13 CSPs, 3 GRs, 4IRs, 25 ORs, and 4 SNMPs in the parasitoid wasps of fruit flies fed with different fruits.

3.7. qRT-PCR

We further used qRT-PCR to validate the expression levels of three randomly selected transcripts in the parasitoid wasps. The primers for the three transcripts and internal control (actin) were predicted using Prime3 and can be accessed in Table 4. We used Log2FC and RNE to show the expression changes of the transcripts in C, M, and G identified by RNA-Seq and qRT-PCR, respectively. Overall, the expression patterns of these transcripts in most comparisons were consistent by both RNA-Seq and qRT-PCR except three (TRINITY_DN1020_c0_g1_i3, TRINITY_DN500_c0_g1_i2 in C_vs_M, and TRINITY_DN1020_c0_g1_i3 in M_vs_G). The high agreement of the gene expression patterns in RNA-Seq and qRT-PCR indicates that the transcripts identified in this study might be functionally expressed in the parasitoid wasps maintained with fruit flies with different scents, which requires future functional experiments.

Table 4. qRT-PCR validation. Log_2 FC represents the log_2 values of the fold change of a transcript identified by the RNA-Seq while RNE represents the relative normalized expression of a transcript identified by qRT-PCR.

		TRINITY_DN1020_c0_g1_i3	TRINITY_DN1284_c0_g1_i11	TRINITY_DN500_c0_g1_i2
Primers	Forward	CAACTTCAAGAACAATCCGACAAC	ACTTATAGACGCATGCCAAGACC	GACGTCGCTATGAACGCTTG
	Reverse	CCACAGCCAGAGACACAGC	GGGCTGCAGAACGGGGATG	GATTCTGATTTCCAGTACGAATACG
C_vs_G	Log_2 FC	10.00	9.84	0
	p-value	4.12×10^{-28}	6.02×10^{-27}	1
	RNE	1.13	2.32	0.42
C_vs_M	Log_2 FC	−9.47	1.43	6.43
	p-value	2.28×10^{-23}	0.0013	1.98×10^{-33}
	RNE	1.68	0.66	0.00
M_vs_G	Log_2 FC	−0.53	8.32	6.66
	p-value	0.3856	1.13×10^{-15}	7.18×10^{-34}
	RNE	0.67	3.52	79.36

4. Discussion

This project was initiated in an effort to identify the olfactory proteins of parasitoid wasps. In recent years, many aspects about the perception of pheromones and other odorants have been elucidated [36]. Olfaction is used by insects to recognize volatile cues that allow the detection of food, predators, and mates [37]. We identified 43 OBPs, 69 CSPs, 60 GRs, 689 ORs, 26 IRs, and 14 SNMPs in *D. longicaudata* (Table 2, Table S2) and some of them were differentially expressed when they were maintained with fruit flies fed with different fruits (Figure 4C, Table S4). This is the first time the chemosensory genes in *D. longicaudata* have been investigated and our findings provide a basis for elucidating the molecular mechanisms of the olfactory-related behaviors of parasitoid wasps.

OBPs are a group of proteins that specialize in the transport of lipids. In this study, we identified 43 transcripts encoding OBPs and this number is similar to the number of genes encoding OBPs in the *Drosophila* genome [38]. Addittionally, classic OBPs have been reported to contain six Cys in their sequences [38,39]. Because all the identified OBPs in this study were general OBPs (GOBPs) (Table S2), we found conserved Cys in their sequences (Figure S1), which increased the confidence of using the transcriptome to identify the OBPs in *D. longicaudata*. In detail, the *D. longicaudata* OBPs were grouped into four subfamilies: OBP-56, -69, -72, and -83 (Table S2). However, no studies have been demonstrated to distinguish their functions. In general, according to their names, they are called GOBPs because they bind general odorants that are likely to be represented by the same volatiles for most of the species [40]. We also showed the differential expression of nine transcripts encoding OBPs in parasitoid wasps in response to the fruit flies with different scents (Figure 4, Table S4). This might indicate that some specific OBPs are expressed to discriminate different scents.

Similar to OBPs, CSPs are another group of proteins that mediate the olfactory recognition in insects [41]. They are thought to be expressed during nearly the whole life circle of insects [42,43]. The number of CSP genes varies in species. For example, only 4~8 CSP genes in ants, flies, bees, wasps, and anopheline mosquitoes [44]; 19~20 in butterfly, moth, and beetle [45–47]; and 27 to 83 in *Culex* mosquito species [48]. We identified 69 CSP transcripts produced by 58 genes (Table S2), the number of which is similar to *Culex* mosquito species. CSPs function not only in the chemical communication between insects and the environment but also in some other cellular processes, such as lipid transport, general immunity, insecticide resistance, and xenobiotic degradation [45,49]. In ants, CSPs have been proposed to mediate the recognition of chemical signatures composed of cuticular lipids [50]. The differential expression of CSP transcripts identified in this study (Table S4) may support the ability to recognize different scents from their hosts.

In the present study, we also identified some other olfactory gene families, such as OR, IR, GR, and SNMP, which were differentially expressed in response to the fruit fly with different scents (Table S2, Table S4). OR is the name for all molecules that are expressed in the cell membranes of olfactory receptor neurons and are responsible for the detection of odorants. The ORs form a multigene family consisting of around 800 genes in humans and 1400 genes in mice [51]. We identified 689 OR transcripts derived from 637 genes (Table 2, Table S2). The diversity of ORs might help insects to discriminate as many different odors as possible. GRs are found be expressed exclusively in gustatory receptor neurons [52]. However, many GRs are not related to taste receptors but function in the detection of sugars, bitter compounds, and non-volatile pheromones [53]. Interestingly, GR28B represents a new class of thermosensor and is required for thermotaxis [54]. We found GR28B differentially expressed in C and M (Figure 4C, Table S2). This might be evidence of its new role in the detection of different scents. IR genes are expressed in coeloconic sensilla of the antenna and respond, among others, to water and amines [55]. IRs are not related to insect ORs but rather have evolved from ionotropic glutamate receptors (iGluRs), a conserved family of synaptic ligand-gated ion channels [56]. In this study, we identified 26 IR transcripts, and 18 of them encode IR25A (Table 2, Table S2). It is conceivable that the IR25A ancestor initially evolved as a sensory detector for external glutamate, analogous to the synaptic function of iGluRs, and that it only later acquired a co-receptor function after duplication and diversification of the IR repertoire [56]. We also identified 14 SNMP transcripts in the parasitoid wasps, including 11 encoding SNMP1 and 3 encoding SNMP2 (Table 2, Table S2). The SNMP1 has been shown to be antenna specific and play an important role in pheromone detection [57]. While SNMP2, which acts as a second lepidoperan and also associates with pheromone-sensitive sensilla, has been shown to be expressed in sensilla support cells rather than neurons [58,59]. The identification and differential expression of olfactory-related transcripts revealed the complex chemosensory system of *D. longicaudata* and supported a diverse function of olfactory genes in discriminating different chemical cues.

The limitations of this project may include the use of the tissue mixture of insects. It is said that some OBPs are expressed in insects with a sex preference. For example, MsepOBP5 exhibited female-biased expression in 0- and 5-day-old adults; MsepOBP22 displayed female-biased expression in 0- and 5-day-old adults but was male-biased in 3-day-old adults [60]. Due to the difficulty of the sample preparation, it is hard to get enough material for sequencing with the same sex. Additionally, it is difficult to determine the tissue-specific olfactory genes. However, our findings provide a basis of future studies about the olfactory system in *D. longicaudata*.

5. Conclusions

In conclusion, we assembled the first transcriptome for *D. longicaudata* using transcriptome sequencing and identified 43 OBPs, 69 CSPs, 60 GRs, 689 ORs, 26 IRs, and 14 SNMPs. Further, 9 OBPs, 13 CSPs, 3 GRs, 4IRs, 25 ORs, and 4 SNMPs were differentially expressed in the insects in response to fruit flies with different scents. Our findings provide a basis towards understanding the molecular mechanisms of *D. longicaudata* in the detection of chemosensory cues. Additionally, the sequences including olfactory genes, noncoding genes, and miRNAs identified in this study can be used in the future and benefit other researchers in this field.

Supplementary Materials: The following are available online at http://www.mdpi.com/2073-4425/11/2/144/s1, Figure S1: Multiple sequence alignment of OBP sequences identified in this study. It showed the conserved six Cys (C1~C6) in these sequences (highlighted in yellow and marked with *). Table S1: miRNA precursors identified in the parasitoid wasp transcriptome. Table S2: Expression profile of transcripts encoding olfactory proteins. Table S3: OBP and CSP sequences for the phylogenetic analysis. Table S4: Differentially expressed transcripts in the parasitoid wasps of fruit flies fed with different fruits.

Author Contributions: Conceptualization, H.Z. and W.L.; methodology, L.T. and J.L.; software, Y.Y.; validation, L.T., J.L. and H.Z.; formal analysis, L.T.; investigation, H.Z.; resources, J.L. and L.L.; data curation, J.L. and Y.Y.; writing—original draft preparation, L.T. and J.L.; writing—review and editing, H.Z. and W.L.; visualization, J.L.; supervision, H.Z.; project administration, H.Z.; funding acquisition, H.Z. All authors have read and agreed to the published version of the manuscript.

Funding: This research was funded by the National Natural Science Foundation of China (31560531) and Guangxi Innovation-driven Projects (AA17202017-2).

Acknowledgments: We acknowledge all the members from the Key Laboratory of Intergrated Pest Management on Tropical Crops, Ministry of Agriculture, Environment and Plant Protection Institute, Chinese Academy of Tropical Agricultural Sciences for their helpful discussions and kindness opinions.

Conflicts of Interest: The authors declare no conflict of interest.

References

1. White, I.M.; Elson-Harris, M.M. *Fruit Flies of Economic Significance: Their Identification and Bionomics*; CAB International: Wallingford, UK, 1992.
2. Lawrence, P.O. Host vibration—A cue to host location by the parasite, Biosteres longicaudatus. *Oecologia* **1981**, *48*, 249–251. [CrossRef]
3. Greany, P.D.; Tumlinson, J.H.; Chambers, D.L.; Boush, G.M. Chemically mediated host finding byBiosteres (Opius) longicaudatus, a parasitoid of tephritid fruit fly larvae. *J. Chem. Ecol.* **1977**, *3*, 189–195. [CrossRef]
4. Carrasco, M.; Montoya, P.; Cruz-lopez, L.; Rojas, J.C. Response of the Fruit Fly Parasitoid Diachasmimorpha longicaudata (Hymenoptera: Braconidae) to Mango Fruit Volatiles. *Environ. Entomol.* **2005**, *34*, 576–583. [CrossRef]
5. Silva, J.W.P.; Bento, J.M.S.; Zucchi, R.A. Olfactory response of three parasitoid species (Hymenoptera: Braconidae) to volatiles of guavas infested or not with fruit fly larvae (Diptera: Tephritidae). *Biol. Control.* **2007**, *41*, 304–311. [CrossRef]
6. Duan, J.J.; Messing, R.H. Effects of host substrate and vibration cues on ovipositor-probing behavior in two larval parasitoids of tephritid fruit flies. *J. Insect Behav.* **2000**, *13*, 175–186. [CrossRef]
7. Stuhl, C.; Sivinski, J.; Teal, P.; Paranhos, B.; Aluja, M. A compound produced by fruigivorous Tephritidae (Diptera) larvae promotes oviposition behavior by the biological control agent Diachasmimorpha longicaudata (Hymenoptera: Braconidae). *Environ. Entomol.* **2011**, *40*, 727–736. [CrossRef]
8. De Bruyne, M.; Warr, C.G. Molecular and cellular organization of insect chemosensory neurons. *Bioessays* **2006**, *28*, 23–34. [CrossRef]
9. Zubkov, S.; Gronenborn, A.M.; Byeon, I.J.; Mohanty, S. Structural consequences of the pH-induced conformational switch in *A. polyphemus* pheromone-binding protein: Mechanisms of ligand release. *J. Mol. Biol.* **2005**, *354*, 1081–1090. [CrossRef]
10. Vogt, R.G.; Riddiford, L.M.; Prestwich, G.D. Kinetic properties of a sex pheromone-degrading enzyme: the sensillar esterase of *Antheraea polyphemus*. *Proc. Natl. Acad. Sci. USA* **1985**, *82*, 8827–8831. [CrossRef]
11. Hull, J.J.; Perera, O.P.; Snodgrass, G.L. Cloning and expression profiling of odorant-binding proteins in the tarnished plant bug, *Lygus lineolaris*. *Insect Mol. Biol.* **2014**, *23*, 78–97. [CrossRef]
12. Sanchez-Gracia, A.; Vieira, F.G.; Rozas, J. Molecular evolution of the major chemosensory gene families in insects. *Heredity (Edinb.)* **2009**, *103*, 208–216. [CrossRef] [PubMed]
13. Jin, F.; Dong, X.; Xu, X.; Ren, S. cDNA cloning and recombinant expression of the general odorant binding protein II from *Spodoptera litura*. *Sci. China C Life Sci.* **2009**, *52*, 80–87. [CrossRef] [PubMed]
14. Xu, P.; Wang, Y.; Akami, M.; Niu, C.Y. Identification of olfactory genes and functional analysis of BminCSP and BminOBP21 in *Bactrocera minax*. *PLoS ONE* **2019**, *14*, e0222193. [CrossRef] [PubMed]
15. Zhu, J.; Yang, P.; Wu, G.; He, Q.; Yin, L.; Xiong, Z. Clone and sequence analysis of the odorant-binding protein genes in *Nasonia vitripennis*. *J. Environ. Entomol.* **2010**, *32*, 476–482.
16. Zhang, K. *Characterization of MmedOBP8 and Its Binding Capacity in Microplitis mediator (Halidag)*; Huazhong Agricultural University: Wuhan, China, 2011.
17. Zhao, Y.; Wang, F.; Zhang, X.; Zhang, S.; Guo, S.; Zhu, G.; Liu, Q.; Li, M. Transcriptome and Expression Patterns of Chemosensory Genes in Antennae of the Parasitoid Wasp *Chouioia cunea*. *PLoS ONE* **2016**, *11*, e0148159. [CrossRef]
18. Harbi, A.; de Pedro, L.; Ferrara, F.A.A.; Tormos, J.; Chermiti, B.; Beitia, F.; Sabater-Munoz, B. *Diachasmimorpha longicaudata* Parasitism Response to Medfly Host Fruit and Fruit Infestation Age. *Insects* **2019**, *10*, 211. [CrossRef]
19. Chen, M.; Xu, R.; Rai, A.; Suwakulsiri, W.; Izumikawa, K.; Ishikawa, H.; Greening, D.W.; Takahashi, N.; Simpson, R.J. Distinct shed microvesicle and exosome microRNA signatures reveal diagnostic markers for colorectal cancer. *PLoS ONE* **2019**, *14*, e0210003. [CrossRef]

20. Chen, M.; Xu, R.; Ji, H.; Greening, D.W.; Rai, A.; Izumikawa, K.; Ishikawa, H.; Takahashi, N.; Simpson, R.J. Transcriptome and long noncoding RNA sequencing of three extracellular vesicle subtypes released from the human colon cancer LIM1863 cell line. *Sci. Rep.* **2016**, *6*, 38397. [CrossRef]
21. Wei, S.; Ma, X.; Pan, L.; Miao, J.; Fu, J.; Bai, L.; Zhang, Z.; Guan, Y.; Mo, C.; Huang, H.; et al. Transcriptome Analysis of *Taxillusi chinensis* (DC.) Danser Seeds in Response to Water Loss. *PLoS ONE* **2017**, *12*, e0169177. [CrossRef]
22. Fu, L.; Niu, B.; Zhu, Z.; Wu, S.; Li, W. CD-HIT: Accelerated for clustering the next-generation sequencing data. *Bioinformatics* **2012**, *28*, 3150–3152. [CrossRef]
23. Seppey, M.; Manni, M.; Zdobnov, E.M. BUSCO: Assessing Genome Assembly and Annotation Completeness. *Methods Mol. Biol.* **2019**, *1962*, 227–245. [CrossRef] [PubMed]
24. Conesa, A.; Gotz, S.; Garcia-Gomez, J.M.; Terol, J.; Talon, M.; Robles, M. Blast2GO: A universal tool for annotation, visualization and analysis in functional genomics research. *Bioinformatics* **2005**, *21*, 3674–3676. [CrossRef]
25. Lagesen, K.; Hallin, P.; Rodland, E.A.; Staerfeldt, H.H.; Rognes, T.; Ussery, D.W. RNAmmer: Consistent and rapid annotation of ribosomal RNA genes. *Nucleic Acids Res.* **2007**, *35*, 3100–3108. [CrossRef]
26. Finn, R.D.; Clements, J.; Eddy, S.R. HMMER web server: Interactive sequence similarity searching. *Nucleic Acids Res.* **2011**, *39*, W29–W37. [CrossRef] [PubMed]
27. Petersen, T.N.; Brunak, S.; von Heijne, G.; Nielsen, H. SignalP 4.0: Discriminating signal peptides from transmembrane regions. *Nat. Methods* **2011**, *8*, 785–786. [CrossRef] [PubMed]
28. Krogh, A.; Larsson, B.; von Heijne, G.; Sonnhammer, E.L. Predicting transmembrane protein topology with a hidden Markov model: Application to complete genomes. *J. Mol. Biol.* **2001**, *305*, 567–580. [CrossRef] [PubMed]
29. Powell, S.; Forslund, K.; Szklarczyk, D.; Trachana, K.; Roth, A.; Huerta-Cepas, J.; Gabaldon, T.; Rattei, T.; Creevey, C.; Kuhn, M.; et al. eggNOG v4.0: Nested orthology inference across 3686 organisms. *Nucleic Acids Res.* **2014**, *42*, D231–D239. [CrossRef]
30. Kong, L.; Zhang, Y.; Ye, Z.Q.; Liu, X.Q.; Zhao, S.Q.; Wei, L.; Gao, G. CPC: Assess the protein-coding potential of transcripts using sequence features and support vector machine. *Nucleic Acids Res.* **2007**, *35*, W345–349. [CrossRef]
31. Li, R.; Yu, C.; Li, Y.; Lam, T.W.; Yiu, S.M.; Kristiansen, K.; Wang, J. SOAP2: An improved ultrafast tool for short read alignment. *Bioinformatics* **2009**, *25*, 1966–1967. [CrossRef]
32. John, B.; Enright, A.J.; Aravin, A.; Tuschl, T.; Sander, C.; Marks, D.S. Human MicroRNA targets. *PLoS Biol.* **2004**, *2*, e363. [CrossRef]
33. Li, B.; Dewey, C.N. RSEM: Accurate transcript quantification from RNA-Seq data with or without a reference genome. *BMC Bioinform.* **2011**, *12*, 323. [CrossRef] [PubMed]
34. Robinson, M.D.; McCarthy, D.J.; Smyth, G.K. edgeR: A Bioconductor package for differential expression analysis of digital gene expression data. *Bioinformatics* **2010**, *26*, 139–140. [CrossRef] [PubMed]
35. Kumar, S.; Stecher, G.; Li, M.; Knyaz, C.; Tamura, K. MEGA X: Molecular Evolutionary Genetics Analysis across Computing Platforms. *Mol. Biol. Evol.* **2018**, *35*, 1547–1549. [CrossRef] [PubMed]
36. Zhou, J.J. Odorant-binding proteins in insects. *Vitam. Horm.* **2010**, *83*, 241–272. [CrossRef] [PubMed]
37. Zhou, S.S.; Sun, Z.; Ma, W.; Chen, W.; Wang, M.Q. De novo analysis of the *Nilaparvata lugens* (Stal) antenna transcriptome and expression patterns of olfactory genes. *Comp. Biochem. Physiol. Part D Genomics Proteomics* **2014**, *9*, 31–39. [CrossRef] [PubMed]
38. Hekmat-Scafe, D.S.; Scafe, C.R.; McKinney, A.J.; Tanouye, M.A. Genome-wide analysis of the odorant-binding protein gene family in *Drosophila melanogaster*. *Genome Res.* **2002**, *12*, 1357–1369. [CrossRef]
39. Gong, M.; Jiao, L.; Ma, W. Orthogonal immune algorithm with diversity-based selection for numerical optimization. In Proceedings of the First ACM/SIGEVO Summit on Genetic and Evolutionary Computation, Shanghai, China, 12–14 June 2009; pp. 141–148.
40. Vogt, R.G.; Prestwich, G.D.; Lerner, M.R. Odorant-binding-protein subfamilies associate with distinct classes of olfactory receptor neurons in insects. *J. Neurobiol.* **1991**, *22*, 74–84. [CrossRef]
41. Vogt, R.G.; Riddiford, L.M. Pheromone binding and inactivation by moth antennae. *Nature* **1981**, *293*, 161–163. [CrossRef]

42. Picimbon, J.F.; Dietrich, K.; Angeli, S.; Scaloni, A.; Krieger, J.; Breer, H.; Pelosi, P. Purification and molecular cloning of chemosensory proteins from *Bombyx mori*. *Arch. Insect. Biochem. Physiol.* **2000**, *44*, 120–129. [CrossRef]
43. Ma, C.; Cui, S.; Tian, Z.; Zhang, Y.; Chen, G.; Gao, X.; Tian, Z.; Chen, H.; Guo, J.; Zhou, Z. OcomCSP12, a Chemosensory Protein Expressed Specifically by Ovary, Mediates Reproduction in *Ophraella communa* (Coleoptera: Chrysomelidae). *Front. Physiol.* **2019**, *10*, 1290. [CrossRef]
44. Liu, G.; Yue, S.; Rajashekar, B.; Picimbon, J.-F. Expression of chemosensory protein (CSP) structures in *Pediculus humanis* corporis and Acinetobacter *A. baumannii*. *SOJ Microbiol. Infect. Dis.* **2019**, *7*, 1–17.
45. Xuan, N.; Guo, X.; Xie, H.Y.; Lou, Q.N.; Lu, X.B.; Liu, G.X.; Picimbon, J.F. Increased expression of CSP and CYP genes in adult silkworm females exposed to avermectins. *Insect. Sci.* **2015**, *22*, 203–219. [CrossRef] [PubMed]
46. Ozaki, K.; Utoguchi, A.; Yamada, A.; Yoshikawa, H. Identification and genomic structure of chemosensory proteins (CSP) and odorant binding proteins (OBP) genes expressed in foreleg tarsi of the swallowtail butterfly *Papilio xuthus*. *Insect. Biochem. Mol. Biol.* **2008**, *38*, 969–976. [CrossRef] [PubMed]
47. Liu, G.; Arnaud, P.; Offmann, B.; Picimbon, J.F. Genotyping and Bio-Sensing Chemosensory Proteins in Insects. *Sensors (Basel)* **2017**, *17*, 1801. [CrossRef]
48. Mei, T.; Fu, W.B.; Li, B.; He, Z.B.; Chen, B. Comparative genomics of chemosensory protein genes (CSPs) in twenty-two mosquito species (Diptera: Culicidae): Identification, characterization, and evolution. *PLoS ONE* **2018**, *13*, e0190412. [CrossRef]
49. Liu, G.; Ma, H.; Xie, H.; Xuan, N.; Guo, X.; Fan, Z.; Rajashekar, B.; Arnaud, P.; Offmann, B.; Picimbon, J.F. Biotype Characterization, Developmental Profiling, Insecticide Response and Binding Property of *Bemisia tabaci* Chemosensory Proteins: Role of CSP in Insect Defense. *PLoS ONE* **2016**, *11*, e0154706. [CrossRef]
50. Ozaki, M.; Wada-Katsumata, A.; Fujikawa, K.; Iwasaki, M.; Yokohari, F.; Satoji, Y.; Nisimura, T.; Yamaoka, R. Ant nestmate and non-nestmate discrimination by a chemosensory sensillum. *Science* **2005**, *309*, 311–314. [CrossRef]
51. Niimura, Y. Evolutionary dynamics of olfactory receptor genes in chordates: Interaction between environments and genomic contents. *Hum. Genomics* **2009**, *4*, 107–118. [CrossRef]
52. Montell, C. Gustatory receptors: Not just for good taste. *Curr. Biol.* **2013**, *23*, R929–R932. [CrossRef]
53. Montell, C. A taste of the Drosophila gustatory receptors. *Curr. Opin. Neurobiol.* **2009**, *19*, 345–353. [CrossRef]
54. Ni, L.; Bronk, P.; Chang, E.C.; Lowell, A.M.; Flam, J.O.; Panzano, V.C.; Theobald, D.L.; Griffith, L.C.; Garrity, P.A. A gustatory receptor paralogue controls rapid warmth avoidance in Drosophila. *Nature* **2013**, *500*, 580–584. [CrossRef]
55. Benton, R.; Vannice, K.S.; Gomez-Diaz, C.; Vosshall, L.B. Variant ionotropic glutamate receptors as chemosensory receptors in Drosophila. *Cell* **2009**, *136*, 149–162. [CrossRef] [PubMed]
56. Rytz, R.; Croset, V.; Benton, R. Ionotropic receptors (IRs): Chemosensory ionotropic glutamate receptors in Drosophila and beyond. *Insect Biochem. Mol. Biol.* **2013**, *43*, 888–897. [CrossRef] [PubMed]
57. Vogt, R.G.; Miller, N.E.; Litvack, R.; Fandino, R.A.; Sparks, J.; Staples, J.; Friedman, R.; Dickens, J.C. The insect SNMP gene family. *Insect. Biochem. Mol. Biol.* **2009**, *39*, 448–456. [CrossRef] [PubMed]
58. Rogers, M.E.; Krieger, J.; Vogt, R.G. Antennal SNMPs (sensory neuron membrane proteins) of Lepidoptera define a unique family of invertebrate CD36-like proteins. *J. Neurobiol.* **2001**, *49*, 47–61. [CrossRef] [PubMed]
59. Forstner, M.; Gohl, T.; Gondesen, I.; Raming, K.; Breer, H.; Krieger, J. Differential expression of SNMP-1 and SNMP-2 proteins in pheromone-sensitive hairs of moths. *Chem. Senses* **2008**, *33*, 291–299. [CrossRef] [PubMed]
60. Chang, X.Q.; Nie, X.P.; Zhang, Z.; Zeng, F.F.; Lv, L.; Zhang, S.; Wang, M.Q. De novo analysis of the oriental armyworm *Mythimna separata* antennal transcriptome and expression patterns of odorant-binding proteins. *Comp. Biochem. Physiol Part. D Genom. Proteom.* **2017**, *22*, 120–130. [CrossRef]

Article

Transcriptome Analysis of Zebrafish Olfactory Epithelium Reveal Sexual Differences in Odorant Detection

Ying Wang [1], Haifeng Jiang [2,3] and Liandong Yang [2,*]

1 Hubei Engineering Research Center for Protection and Utilization of Special Biological Resources in the Hanjiang River Basin, School of Life Sciences, Jianghan University, Wuhan 430056, Hubei, China; wangying1987@jhun.edu.cn

2 The Key Laboratory of Aquatic Biodiversity and Conservation of Chinese Academy of Sciences, Institute of Hydrobiology, Chinese Academy of Sciences, Wuhan 430072, Hubei, China; jianghf@ihb.ac.cn

3 University of Chinese Academy of Sciences, Beijing 100049, China

* Correspondence: yangld@ihb.ac.cn; Tel.: +86-27-6878-0281

Received: 3 March 2020; Accepted: 25 May 2020; Published: 27 May 2020

Abstract: Animals have evolved a large number of olfactory receptor genes in their genome to detect numerous odorants in their surrounding environments. However, we still know little about whether males and females possess the same abilities to sense odorants, especially in fish. In this study, we used deep RNA sequencing to examine the difference of transcriptome between male and female zebrafish olfactory epithelia. We found that the olfactory transcriptomes between males and females are highly similar. We also found evidence of some genes showing differential expression or alternative splicing, which may be associated with odorant-sensing between sexes. Most chemosensory receptor genes showed evidence of expression in the zebrafish olfactory epithelium, with a higher expression level in males than in females. Taken together, our results provide a comprehensive catalog of the genes mediating olfactory perception and pheromone-evoked behavior in fishes.

Keywords: zebrafish; sexual dimorphism; olfactory epithelium; alternative splicing; chemosensory receptor

---

1. Introduction

Olfaction is a sense for detecting environmental odorants that plays essential roles in many aspects of animal activities, such as foraging, migration, prey avoidance and mating [1]. In most terrestrial vertebrates, there are two distinct chemosensory organs: the olfactory epithelium (OE) and the vomeronasal organ (VNO) [2]. In general, the OE is assumed to detect environmental odorants, while the VNO senses pheromones, although some exceptions were also reported in recent studies [3,4]. Different set of chemosensory receptors is thought to be expressed in each organ. For the OE, odorant receptors (ORs) and trace amine-associated receptors (TAARs) are known to be expressed, while for the VNO, vomeronasal receptors type 1 (V1Rs) and type 2 (V2Rs) are considered to be expressed [5]. Therefore, in terrestrial mammals, ORs and TAARs are suggested to detect "ordinary" odorant molecules, while V1Rs and V2Rs are employed to detect pheromones [6–8].

However, unlike the terrestrial counterparts, fish have only the OE to sense their olfactory environment, due to an absence of the VNO [9]. Without directly functional studies of the chemosensory receptors in fish, it is unclear whether the different types of chemosensory receptors in fish respond to different classes of odorants—and which family of the chemosensory receptors is used for recognizing odorants or pheromones. Interestingly, there are also some studies using electrophysiological experiments that provide indirect evidence of putative ligands for chemosensory receptors [10–16]. For example, members of ORs from goldfish can recognize F-prostaglandins, while both ORs and V2Rs can perceive amino acids [17,18].

Several studies have found that sex-specific behaviors can be due to sex differences in responses to external stimuli, including courtship songs, colors and chemosensory cues [19]. Indeed, chemosensory cues in mice, such as pheromone molecules, have been shown to be detected specifically by their vomeronasal receptors [20]. Therefore, sex-specific behaviors maybe initiated by sex-specific pheromones [21–23] or by sex-differential expression of chemosensory receptors in response to the same pheromones [24]—or even by sex-specific alternative splicing [25]. For instance, Darcin (the MUP20 encoded by Mup20)—a male-specific pheromone in mouse urine—only attracts females to affect their memory [24]. ESP1 (exocrine-secreted peptide 1) is detected by both male and female mice, but only stimulated female-specific mating behaviors, suggesting the existence of sex-specific neuronal circuits [23,26]. Male fruit flies require the protein encoded by the fruitless (fru) gene to complete courtship, which is produced in different sex-specific isoforms via alternative splicing [25]. However, to date, few studies have reported whether sexual differences in odorant-sensing exists between males and females, especially in fish.

To determine the full repertoire of chemosensory receptors expressed in the zebrafish olfactory epithelium and evaluate the extent of sexual differentiation in odorant-sensing between sexes, we used RNAseq to profile their transcriptomes in male and female zebrafish. We found that a very high percentage of chemosensory receptors are indeed expressed in the zebrafish olfactory epithelium, with a higher expression level in males than females. However, the olfactory transcriptomes between males and females are highly similar, with limited genes showing differentially expressed or alternatively splicing, which may be associated with odorant-sensing between sexes. Collectively, our results provide a comprehensive catalog of the genes mediating olfactory perception and pheromone-evoked behavior in fish.

2. Materials and Methods

2.1. Sample Collection

All experiments in this research were approved by the Institutional Animal Care and Use Committee of Institute of Hydrobiology, Chinese Academy of Sciences (Approval ID: Y21304501). Adult wild-type zebrafish from AB background were maintained in the zebrafish facilities in the China Zebrafish Resource Center (CZRC)for one week to familiarize with the laboratory environment. Zebrafish were raised together in a mixed sex population. Olfactory epithelium from each individual was dissected out and frozen in liquid nitrogen quickly. Due to their small size, each of the samples was a pool of mRNA from three individuals of the same sex. Three independent biologic replicates for both male and female samples were prepared.

2.2. Library Construction and High-Throughput Sequencing

Total RNA from each of the six samples was extracted using the SV Total RNA Isolation System (Promega). We assessed the RNA quality using agarose gel electrophoresis and measured RNA integrity using the RNA Nano 6000 Assay Kit of the Agilent Bioanalyzer 2100 system (Agilent Technologies, Santa Clara, CA, USA). Sequencing libraries preparation and high throughput sequencing were generated by Novogene (Beijing, China) following our previous study [27–29]. Briefly, mRNA was purified from total RNA with poly-T oligo-attached magnetic beads and fragmented into short pieces. Then, the first-strand cDNA was synthesized using random hexamer primer and second-strand cDNA was then generated. Finally, the paired-end cDNA library was prepared according to the Illumina's protocols and sequenced on Illumina HiSeq 4000 platform (150 bp paired-end) (Illumina, San Diego, CA, USA). The RNA-seq reads were deposited into the National Center for Biotechnology Information (NCBI) Sequence Read Archive database (Accession No. SRP154651, Supplementary Materials Table S1).

2.3. Analysis of RNA-Seq Data

Raw RNA-seq reads were first filtered to delete primer dimers and low-quality bases (Phred quality score lower 20) using the Trim Galore! program (version 0.3.7) (https://www.bioinformatics.babraham.ac.uk/projects/trim_galore/). We only retained paired-end reads from which either end was longer than 50 bp after trimming for subsequent analyses. High quality paired-end reads from each sample were aligned to the transcripts from zebrafish genome annotated by Ensembl (release 97) [30] using Bowtie2 [31];abundances of transcripts(FPKM, Fragments Per Kilobase Million)were estimated using RSEM program (v 1.3.1) [32]. We filtered the genes with FPKM > 1 in at least half of the six samples as transcriptionally active genes for subsequent analyses. The raw read counts for each transcript estimated by RSEM were extracted and then normalized using TMM method to control for differences in sequencing depth among samples, and the differentially expressed transcripts were identified using the edgeR package [33] using a minimal fold change of 2 and an adjusted p value cutoff of 0.05. Full lists of differential expression genes can be found in Supplementary Materials Table S2.

2.4. Gene Ontology Analysis

Overrepresentation of the gene ontology (GO) terms for upregulated genes between male and female zebrafish olfactory epithelium were identified using Gorilla (http://cbl-gorilla.cs.technion.ac.il/) [34], which allows detection functional overrepresentation in a candidate data set against a list of background genes. We set the false discovery rate (FDR) of 0.001 as our cutoff value and conducted separately for each sex.

2.5. Characterization of Alternative Splicing Events (ASEs)

ASEs are divided into five broad categories including skipped exon (SE), alternative 5′ splice site (A5SS), alternative 3′ splice site (A3SS), mutually exclusive exons (MXE) and retained intron (RI) [35]. We used rMATS [36] to detect and count reads that correspond to each of the five types of ASEs. rMATS can identify these ASEs events from a GTF file of annotated transcripts and count the number of reads that correspond to each of the five events described. To identify differential alternative splicing (DAS) events between male and female zebrafish olfactory epithelium, an FDR-adjusted p value less than 0.05 was used as threshold for DAS events (Supplementary Materials Table S6).

2.6. Data Mining for the Chemosensory Receptor Repertoire

We extracted all the sequences from annotated and automatically predicted paralogs for *or*, *taar*, *ora/V1r* and *olfC/V2r* genes from the Ensembl zebrafish genome (GRCz11, release 97). We only considered a gene as a putative chemosensory receptor gene for a given family by checking the candidates position within each chemosensory receptor family clade in a phylogenetic analysis. By using this method, we obtained a total of 170 *or*, 126 *taar*, 5*ora/V1r* and 57 *olfC/V2r* genes.

2.7. Phylogenetic Analysis

All of the coding sequences for the chemosensory receptor genes were obtained from Ensembl. The coding sequences for each of the 4 chemosensory receptor gene families were translated into protein sequences, aligned with the program MUSCLE [37] and then back-reversed to their coding sequences alignment. The ML trees were reconstructed by RAxML (version 8.1.17) [38] under the GTRGAMMMAI substitution model with bootstrap support values determined using 1000 replicates.

2.8. Quantitative Real-Time PCR (qRT-PCR)

In order to confirm the differentially expressed genes detected by RNA-seq, we further employed quantitative real-time PCR (qRT-PCR) on a subset of genes that among the significantly DEGs between the male and female zebrafish olfactory epithelium. Primers of these genes were designed using the NCBI primer designing tool (http://www.ncbi.nlm.nih.gov/tools/primer). We synthesized the first strand cDNA from 500 ng of total RNA samples using M-MLV Reverse Transcriptase (Promega, Madison, WI, USA) and diluted 1:10 as amplification template. qRT-PCR was performed in a 10-µL volume using the LightCycler® 480 SYBR Green I Master on a LightCycler® 480 II Instrument (Basel, Roche, Switzerland). Thermocycling conditions were 95 °C for 5 min, followed by 45 cycles of 95 °C for 20 s and 58 °C for 25 s, and a melting curve analysis was performed to confirm the primer specificity after amplification. The relative gene expression between male and female zebrafish olfactory epithelium was determined using the comparative CT method [39] and the fold change values were the mean of six biologic replicates from each group.

3. Results

3.1. Transcriptome Data

A total of six cDNA libraries were constructed in this study. In total, we obtained 23.75, 24.44, 25.65 million reads for male olfactory epithelium and 22.96, 21.70, 22.28 million reads for female olfactory epithelium, respectively. From these, 75.46–77.75% of the reads can be mapped to the annotated regions in zebrafish genome, indicating the good quality of reads (Supplementary Materials Table S1). Overall, we detected a total of 18,658 genes with FPKM > 1 in at least half of the six samples, which were defined as robustly expressed genes. The expression estimates for all annotated transcripts in each replicate were provided in Supplementary Materials Table S2.

We first evaluated the variation in gene expression among the three biologic replicate samples for each sex. We found that the correlation values were highly significant between them all (Figure 1, Pearson correlation coefficients of at least 0.95, p-value $< 2.2 \times 10^{-16}$). Only very small sets of genes are unusually variable among replicates (Figure 1). In general, the OE transcriptomes from the male and the female zebrafish were highly correlated (Figure 2A, Pearson correlation coefficient, $r = 0.89$, $p < 2.2 \times 10^{-16}$). We therefore averaged the FPKM values for each gene across each sex. In the olfactory epithelium from both male and female, a few gene are extremely highly expressed. For example, in male 312 most abundant genes account for almost 50% of the fragments and in female 333 most abundant genes account for almost 50% of the fragments obtained from the tissue (Supplementary Materials Table S3). Moreover, a total of 295 genes were found to be shared between male and female most abundant genes. These results were generally consistent with the patterns found in mouse olfactory transcriptomes [40].

3.2. Sex Dimorphism in Zebrafish Olfactory Epithelium Expression Profiles

To assess whether transcriptional differences in the zebrafish olfactory epithelium can account for sex-specific responses to olfactory cues, we examined their sexually dimorphic gene expression profiles. We found that the overall transcriptomes are highly similar between male and female zebrafish olfactory epithelia (Figure 2A,B). Briefly, we detected a total of 713 transcripts showing higher expression level in male olfactory epithelium and 605 transcripts showing higher expression level in female olfactory epithelium, respectively (Figure 2C). However, only 68 and 53 transcripts remained significantly differential expression between male and female olfactory epithelium after applying a false discovery rate (FDR) threshold of 0.05 (Figure 2D and Supplementary Materials Table S4). Among these, some genes identified to be differentially expressed between sexes are expected to be involved in response to olfactory cues (Table 1), such as gene *OR132-2* (odorant receptor, family H, subfamily 132, member 2), *OTX1* (orthodenticle homeobox 1) and *OR115-10* (odorant receptor, family F, subfamily 115, member 10). However, whether these differentially expressed olfactory receptor genes can detect sex relative pheromones needs to be verified in further functional experiments.

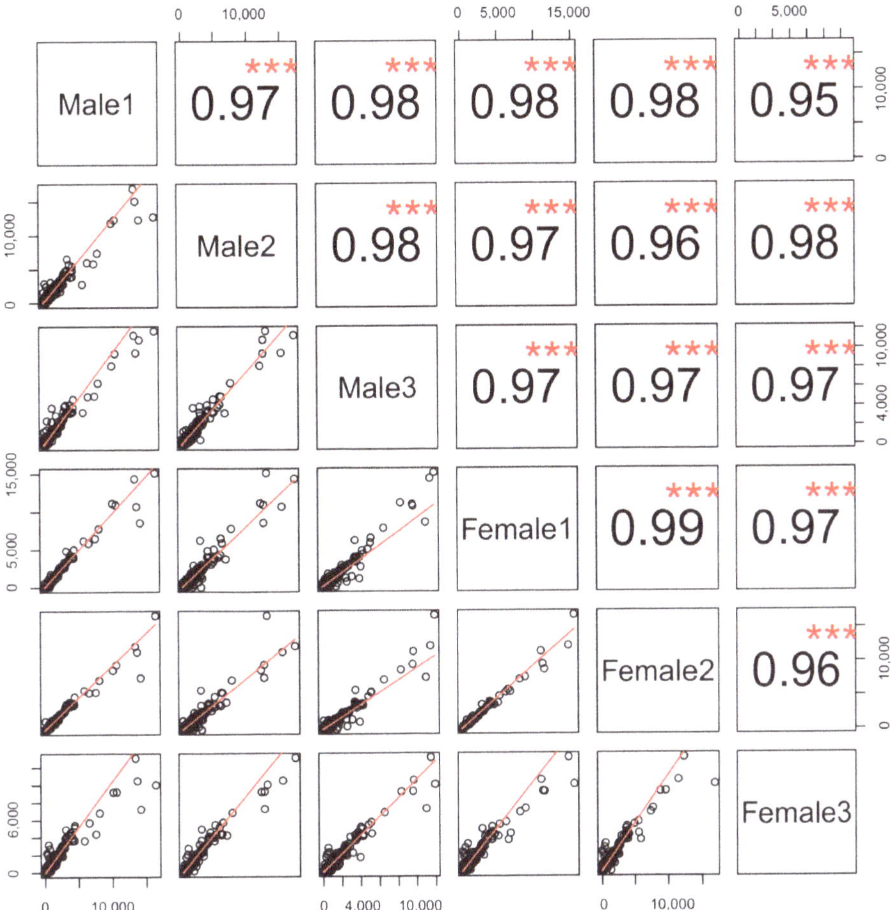

Figure 1. Correlations of gene expression profile between biologic replicates. The three male and three female samples are listed on the diagonal. Above the diagonal the rho value of the Pearson correlation coefficient is indicated. Below are pairwise comparisons between biologic replicates, shown as scatter plots of the abundances of transcripts(FPKM) expression values for all genes. *** means *p*-value > 0.95.

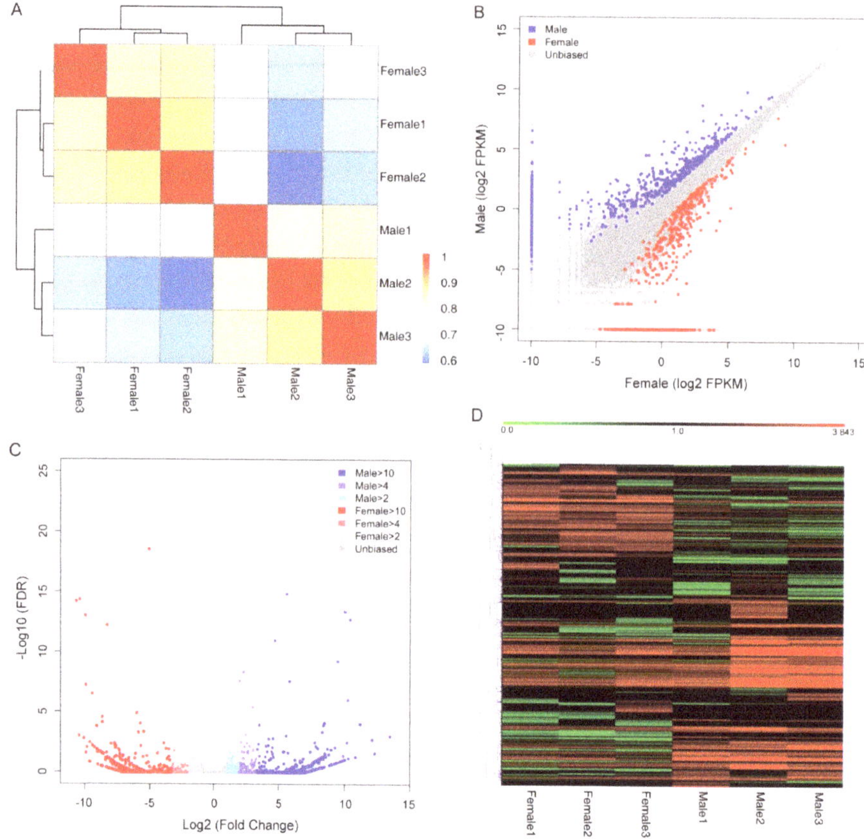

Figure 2. Sexual dimorphism in zebrafish olfactory system. (**A**) Heatmap of cross-correlations of all samples using differentially expressed transcripts; (**B**) pairwise comparison of gene expression abundances between male and female; (**C**) log2-fold change between male and female samples is plotted against their −log10 FDR; (**D**) heatmap of differentially expressed transcripts identified in zebrafish olfactory epithelium.

Table 1. List of selected differentially expressed genes.

Gene ID	Gene Name	Gene Description	Log2(Male/Female)	*p*-Value	FDR
ENSDARG00000105762	or132-2	odorant receptor, family H, subfamily 132, member 2	2.295	1.07×10^{-12}	4.48×10^{-9}
ENSDARG00000094515	or132-2	odorant receptor, family H, subfamily 132, member 2	2.021	7.02×10^{-12}	2.56×10^{-8}
ENSDARG00000035048	or115-10	odorant receptor, family F, subfamily 115, member 10	1.546	3.02×10^{-6}	0.003
ENSDARG00000105835	or132-3	odorant receptor, family H, subfamily 132, member 3	1.606	1.81×10^{-5}	0.012
ENSDARG00000056277	or132-5	odorant receptor, family H, subfamily 132, member 5	1.373	4.86×10^{-5}	0.024
ENSDARG00000105719	or132-4	odorant receptor, family H, subfamily 132, member 4	1.232	5.84×10^{-5}	0.027
ENSDARG00000094992	otx1	orthodenticle homeobox 1	10.205	2.68×10^{-6}	0.003
ENSDARG00000062593	stox1	storkhead box 1	−10.404	2.91×10^{-19}	4.47×10^{-15}
ENSDARG00000075271	rapgef5a	Rap guanine nucleotide exchange factor (GEF) 5a	−8.647	1.13×10^{-8}	2.75×10^{-5}
ENSDARG00000060610	pcdh7b	protocadherin 7b	−8.962	2.51×10^{-5}	0.016

To gain further insights into the biologic processes that differed between male and female zebrafish olfactory epithelium, we performed a GO enrichment analysis with the upregulated genes detected in male and female zebrafish olfactory epithelium (Supplementary Materials Table S5). We found that gene ontology biologic process terms associated with response to external stimulus were highly overrepresented among the differentially expressed genes (DEGs) with male-biased expression (Figure 3 and Supplementary Materials Table S5), including response to other organism (GO:0051707), chemotaxis (GO:0006935) and taxis (GO:0042330). However, a significantly different pattern was found in the enriched GO terms for female olfactory epithelium upregulated genes. For example, female olfactory epithelium upregulated genes were enriched in ion import, such as potassium ion import (GO:0010107), inward rectifier potassium channel activity (GO:0005242) and ligand-gated channel activity (GO:0022834). Taken together, these results indicated that there are significant differences in the upregulated genes between male and female zebrafish olfactory epithelium.

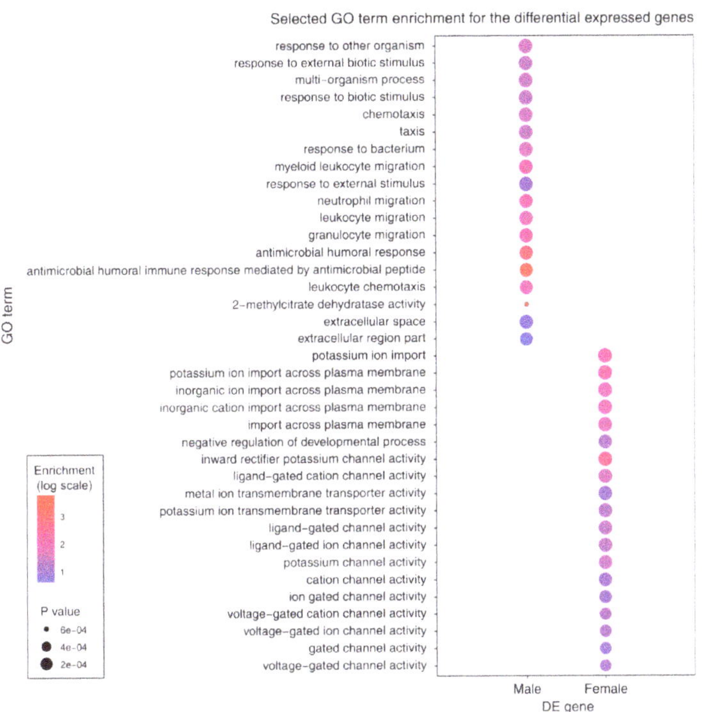

Figure 3. Gene ontology (GO) terms associated with differentially expressed genes between male and female zebrafish olfactory epithelium.

3.3. *Alternative Splicing between the Two Sexes of Zebrafish Olfactory Epithelium*

To examine the influence of sex on alternative splicing regulation in zebrafish olfactory epithelium, alternative splicing events (ASEs) including skipped exon (SE), alternative 5′ splice site (A5SS), alternative 3′ splice site (A3SS), mutually exclusive exons (MXE) and retained intron (RI) (Figure 4A) between male and female zebrafish olfactory epithelium were identified using rMATS [36].

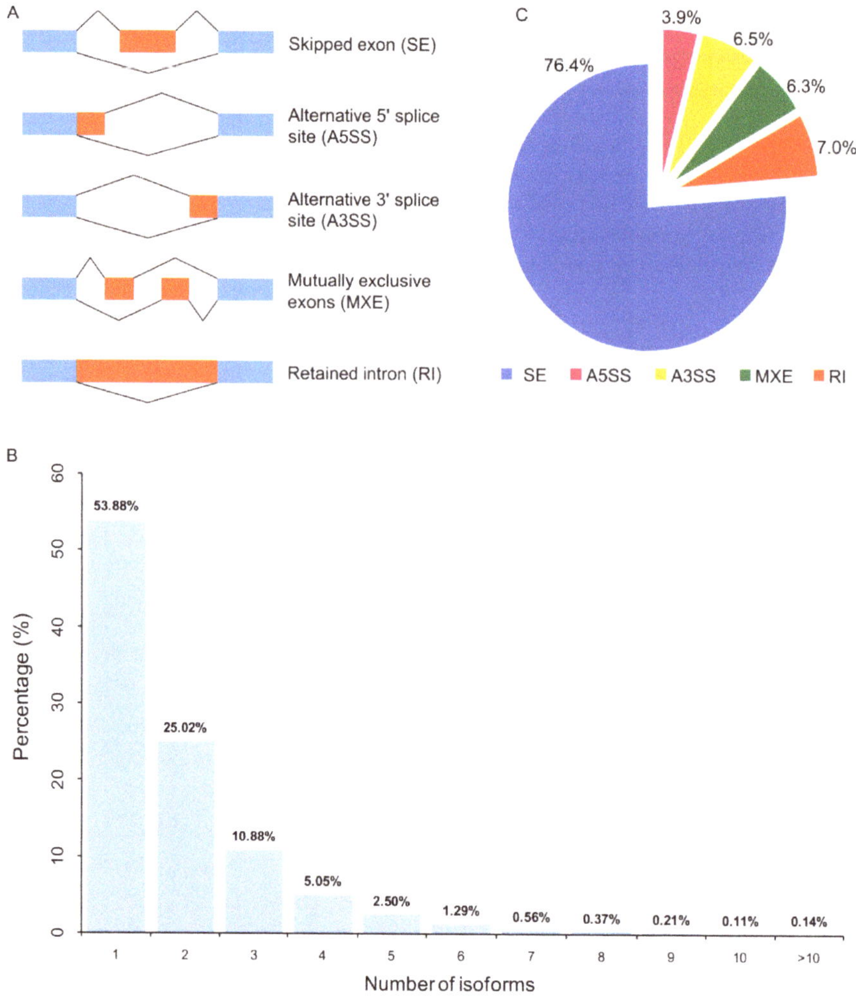

Figure 4. Alternative splicing events (ASEs) between male and female zebrafish olfactory epithelium. (**A**) Schematic representation of five basic types of alternative splicing events. Alternative exons are shown as orange boxes and flanking constitutive exons are shown as blue boxes; (**B**) distribution of isoform numbers for genes in zebrafish genome; (**C**) pie chart showing the percentage distribution of ASEs.

Roughly 46% of genes have 2 detectable isoforms in zebrafish genome (Figure 4B). Among them, we detected a total of 15,193 ASEs, which were distributed in 7585 genes. The most abundant ASEs were skipped exon, accounting for 76.4% of all ASEs, followed by retained intron (7.0%), alternative 3′ splice site (6.5%), mutually exclusive exons (6.3%) and alternative 5′ splice site (3.9%) (Figure 4C). Using p value cutoff of 0.05, differential ASEs were identified between the two sexes of zebrafish (Supplementary Materials Table S6). A total of 86 significantly differential ASEs were identified between the male and female zebrafish olfactory epithelium, which included 50 SE, 4 A5SS, 17 A3SS, 6 MXE, and, 9 RI (Table 2). For example, *maptb*, which is expressed in the developing central nervous system [41], was found to generate different isoforms by sex. *stxbp5l*, which is involved in syntaxin binding, also showed differential isoforms between sexes, which was identified as sexually dimorphic gene between cattle and rat species [42].

Table 2. Of selected differential alternative splicing events.

Gene ID	Gene Name	Gene Description	ASEs Type	p-Value	FDR
ENSDARG00000087616	*maptb*	microtubule-associated protein tau b	SE	1.79×10^{-8}	6.91×10^{-5}
ENSDARG00000006383	*stxbp5l*	syntaxin binding protein 5-like	SE	2.26×10^{-6}	0.002
ENSDARG00000019208	*camsap1a*	calmodulin regulated spectrin-associated protein 1a	SE	0.0001	0.04
ENSDARG00000041736	*hsdl1*	hydroxysteroid dehydrogenase like 1	SE	0.0001	0.03
ENSDARG00000055825	*celsr3*	cadherin, EGF LAG seven-pass G-type receptor 3	A5SS	6.9×10^{-6}	0.004
ENSDARG00000014577	*rhpn2*	rhophilin, Rho GTPase binding protein 2	A5SS	3.55×10^{-5}	0.01
ENSDARG00000074255	*micu3b*	mitochondrial calcium uptake family, member 3b	MXE	1.44×10^{-7}	6.92×10^{-5}
ENSDARG00000077818	*nrg2a*	neuregulin 2a	MXE	0.0002	0.03
ENSDARG00000086103	*slc37a1*	solute carrier family 37 (glucose-6-phosphate transporter), member 1	MXE	0	0
ENSDARG00000025338	*hagh*	hydroxyacylglutathione hydrolase	RI	5.9×10^{-7}	0.0003

3.4. The Chemosensory Receptor Repertoires

As in mouse olfactory sensory neurons (OSNs), each of the OSNs in mature zebrafish randomly expressed only one olfactory receptor [43–45]. Thus, the expression level of any given olfactory receptor within the olfactory epithelium will be low. Consistent with this prediction, among the 170 *or*, 126 *taar*, 5*ora*/*V1r* and 57 *olfC*/*V2r* genes in the zebrafish genome, we detected 147 out of 170 *or*, 104 out of 126 *taar*, 4 out of 5*ora*/*V1r* and 57 out of 57 *olfC*/*V2r* genes expressed in either male or female olfactory epithelium with FPKM > 1, confirming that the chemosensory receptors are mainly expressed in the olfactory epithelium (Supplementary Materials Table S7). However, almost all (90%) of these chemosensory receptor genes were found to be very lowly expressed in the zebrafish olfactory epithelium transcriptomes (FPKM < 15), which is consistent with the results from mice [46].

To further assess whether there are differences in chemosensory receptor gene expression levels between the two sexes in zebrafish, we compared the expression levels of chemosensory receptor genes between male and female zebrafish olfactory epithelium transcriptome. From this analysis, we made the following interesting observations. First, the relative receptor abundance level vary greatly between each member, suggesting that each member of the chemosensory receptor genes was expressed asymmetrically. Some members have high expression levels, whereas other members expressed with very low levels (Figure 5). Second, almost all of the chemosensory receptor genes expressed with no differences between male and female zebrafish olfactory epithelium, with only six *or* genes showing differentially expressed between the two sexes (Supplementary Materials Table S7). Third, all of the 6 *or* genes (Table 1) that showed sexual preference were expressed at a higher level in the male olfactory epithelium than in females, which is generally consistent with the results from mice [40,46]. In order to verify the expression results from RNA-seq, we further employed qRT-PCRto measure relative mRNA levels for a total of 12 candidate genes. Our results demonstrated that the expression patterns for these genes were highly consistent between RNA-seq and qRT-PCR (Figure 5E), suggesting the reliability of our RNA-seq datasets.

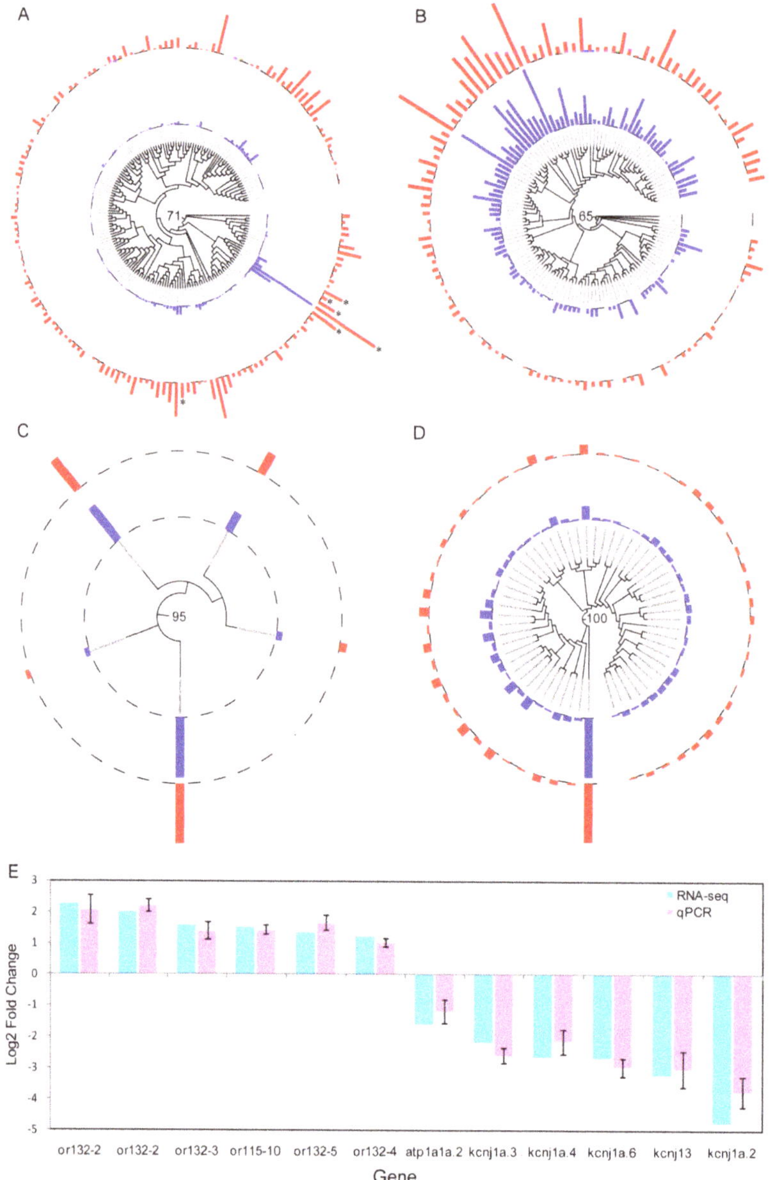

Figure 5. Expression pattern of the chemosensory receptor genes in the zebrafish olfactory epithelium. Mean FPKM expression values across the three samples between male (blue) and female (red) for each of the *or* (**A**), *taar* (**B**), *ora/V1r* (**C**) and *olfC/V2r* (**D**) genes. Phylogenetic trees were reconstructed using RAxML (version 8.1.17) under the GTRGAMMMAI model with bootstrap support values determined using 1000 replicates. * means differentially expressed genes. Bootstrap values for basal nodes are provided; (**E**) validation of RNA-seq data using quantitative real-time PCR (qRT-PCR).

4. Discussion

During the last few decades, chemosensory receptor gene families have received extensive attention across various organisms, including humans [45,47–49], mice [40,46,50], fish [28,51–53] and insects [21,54–57]. However, it remains still unclear whether male and female individual possesses the same ability for odorant-sensing. In the present study, we reported the transcriptional profiles of the olfactory epithelium from the male and female zebrafish obtained by RNA-seq. By comparing the expression levels of genes from three different biologic replicate samples for each sex, we obtained highly correlated samples, which suggested that the subsequent differential expression analyses were reliable.

Sexual dimorphisms in behaviors have been widely observed in the olfactory systems in both mice [50,58] and flies [59]. Although previous studies have suggested that sexually dimorphic behaviors between sexes could be influenced by sensory input such as olfactory cues [60], it is unclear whether differences in gene expression in olfactory epithelia can underlie sexually dimorphic behaviors in fish [40]. Our results showed that the overall transcriptional profiles of zebrafish olfactory epithelium between sexes are highly similar. However, we indeed detected a few genes differentially expressed between male and female zebrafish, whose roles may be associated with odorant-sensing. Moreover, the functional enrichment analyses of the differentially expressed genes were also involved in response to external biotic stimulus. Therefore, all these results from our study suggested that differences in gene expression in the olfactory epithelium between sexes in zebrafish may play a role in odorant sensing to some extent.

However, when we focused on the chemosensory receptor genes that are directly binding to the odorants in surrounding environment, our results suggest that the olfactory receptor genes may not all be involved in behavioral differences between males and females. Although almost all of the chemosensory receptor genes can be detected to be expressed in the olfactory epithelium of zebrafish, most them displayed similar expression levels between sexes, with only six *or* genes showing differentially expressed between the two sexes. Considering the existence of nearly 358 chemosensory receptor genes in zebrafish genome, the six differentially expressed *or* genes should play minor roles in the dimorphic behaviors between male and female zebrafish [61,62]. Therefore, the sexually dimorphic behavioral responses to odorants in zebrafish are unlikely to be solely accounted for by transcriptional differences at the level of detection [40].

Interestingly, several functional studies have identified several olfactory receptor ligands in fish [10–16]. For example, Yabukiet al. reported that *or114-1* and *or114-2* in the group β *or* genes are the key olfactory receptors for the sex pheromone prostaglandin $F_{2\alpha}$ to mediate male courtship behavior in zebrafish [16]. However, none of these olfactory receptor genes were found to be significantly differentially expressed in the olfactory epithelium between male and female in our RNA-seq results. It should be noted that differential expression of peripheral odorant receptors is not the only way to generate sex-specific behavior, as higher neuronal circuits could be different between males and females. Therefore, whether the six differentially expressed *or* genes can play crucial roles in sexually dimorphic behaviors is still uncertain and need to be confirmed by further functional studies in future.

5. Conclusions

In this study, we performed a transcriptomic analysis on the olfactory epithelia of both male and female zebrafish using high-throughput RNA sequencing. We found the olfactory transcriptomes between males and females are highly similar, with only few genes displaying differentially expressed. Most of chemosensory receptor genes showed evidence of expression in the zebrafish olfactory epithelia, with a higher level of expression in males than in females. Collectively, these results provide a comprehensive catalog of the genes mediating olfactory perception and pheromone-evoked behavior in fish.

Supplementary Materials: The following are available online at http://www.mdpi.com/2073-4425/11/6/592/s1, Table S1: Summary of the RNA-seq datasets in olfactory mucosae from male and female zebrafish, Table S2: Expression estimates in the zebrafish olfactory epithelium. A dataset containing the raw and normalized expression values for all the genes in the zebrafish olfactory epithelium, Table S3: 312 and 333 most abundant genes account for 50% of the fragments obtained from the tissue for male and female, Table S4: Differential expression analysis between male and female samples. Genes with an FDR < 0.05 were considered significant, Table S5: Functional enrichment analysis for differentially expressed genes, Table S6: Differential alternative splicing events between male and female zebrafish olfactory epithelium, Table S7 Expression level of the chemosensory receptor genes.

Author Contributions: Conceptualization, L.Y.; formal analysis, Y.W., H.J. and L.Y.; funding acquisition, Y.W. and L.Y.; investigation, H.J.; writing—original draft, Y.W.; writing—review & editing, L.Y. All authors have read and agreed to the published version of the manuscript.

Funding: This research was supported by grants from the National Natural Science Foundation of China (31702016) to Y.W. and from the National Natural Science Foundation of China (31972866) to L.Y.

Acknowledgments: We are grateful to the computational support from the Wuhan Branch, Supercomputing Center, Chinese Academy of Sciences, China.

Conflicts of Interest: The authors declare no conflict of interest.

References

1. Hashiguchi, Y.; Nishida, M. Evolution and origin of vomeronasal-type odorant receptor gene repertoire in fishes. *BMC Evol. Biol.* **2006**, *6*, 76. [CrossRef]
2. Dulac, C.; Torello, A.T. Molecular detection of pheromone signals in mammals: From genes to behaviour. *Nat. Rev. Neurosci.* **2003**, *4*, 551–562. [CrossRef] [PubMed]
3. Baxi, K.N.; Dorries, K.M.; Eisthen, H.L. Is the vomeronasal system really specialized for detecting pheromones? *Trends Neurosci.* **2006**, *29*, 1–7. [CrossRef] [PubMed]
4. Spehr, M.; Kelliher, K.R.; Li, X.H.; Boehm, T.; Leinders-Zufall, T.; Zufall, F. Essential role of the main olfactory system in social recognition of major histocompatibility complex peptide ligands. *J. Neurosci.* **2006**, *26*, 1961–1970. [CrossRef]
5. Mombaerts, P. Genes and ligands for odorant, vomeronasal and taste receptors. *Nat. Rev. Neurosci.* **2004**, *5*, 263–278. [CrossRef]
6. Boschat, C.; Pelofi, C.; Randin, O.; Roppolo, D.; Luscher, C.; Broillet, M.C.; Rodriguez, I. Pheromone detection mediated by a V1r vomeronasal receptor. *Nat. Neurosci.* **2002**, *5*, 1261–1262. [CrossRef]
7. Leinders-Zufall, T.; Brennan, P.; Widmayer, P.; Chandramani, P.; Maul-Pavicic, A.; Jager, M.; Li, X.H.; Breer, H.; Zufall, F.; Boehm, T. MHC class I peptides as chemosensory signals in the vomeronasal organ. *Science* **2004**, *306*, 1033–1037. [CrossRef]
8. Kimoto, H.; Haga, S.; Sato, K.; Touhara, K. Sex-specific peptides from exocrine glands stimulate mouse vomeronasal sensory neurons. *Nature* **2005**, *437*, 898–901. [CrossRef]
9. Grus, W.E.; Zhang, J. Origin and evolution of the vertebrate vomeronasal system viewed through system-specific genes. *Bioessays* **2006**, *28*, 709–718. [CrossRef]
10. Speca, D.J.; Lin, D.M.; Sorensen, P.W.; Isacoff, E.Y.; Ngai, J.; Dittman, A.H. Functional identification of a goldfish odorant receptor. *Neuron* **1999**, *23*, 487–498. [CrossRef]
11. Luu, P.; Acher, F.; Bertrand, H.O.; Fan, J.H.; Ngai, J. Molecular determinants of ligand selectivity in a vertebrate odorant receptor. *J. Neurosci.* **2004**, *24*, 10128–10137. [CrossRef] [PubMed]
12. Oike, H.; Nagai, T.; Furuyama, A.; Okada, S.; Aihara, Y.; Ishimaru, Y.; Marui, T.; Matsumoto, I.; Misaka, T.; Abe, K. Characterization of ligands for fish taste receptors. *J. Neurosci.* **2007**, *27*, 5584–5592. [CrossRef] [PubMed]
13. Hussain, A.; Saraiva, L.R.; Ferrero, D.M.; Ahuja, G.; Krishna, V.S.; Liberles, S.D.; Korsching, S.I. High-affinity olfactory receptor for the death-associated odor cadaverine. *Proc. Natl. Acad. Sci. USA* **2013**, *110*, 19579–19584. [CrossRef]
14. Behrens, M.; Frank, O.; Rawel, H.; Ahuja, G.; Potting, C.; Hofmann, T.; Meyerhof, W.; Korsching, S. ORA1, a Zebrafish Olfactory Receptor Ancestral to All Mammalian V1R Genes, Recognizes 4-Hydroxyphenylacetic Acid, a Putative Reproductive Pheromone. *J. Biol. Chem.* **2014**, *289*, 19778–19788. [CrossRef]
15. Li, Q.; Tachie-Baffour, Y.; Liu, Z.; Baldwin, M.W.; Kruse, A.C.; Liberles, S.D. Non-classical amine recognition evolved in a large clade of olfactory receptors. *Elife* **2015**, *4*, e10441. [CrossRef]

16. Yabuki, Y.; Koide, T.; Miyasaka, N.; Wakisaka, N.; Masuda, M.; Ohkura, M.; Nakai, J.; Tsuge, K.; Tsuchiya, S.; Sugimoto, Y.; et al. Olfactory receptor for prostaglandin F2alpha mediates male fish courtship behavior. *Nat. Neurosci.* **2016**, *19*, 897–904. [CrossRef]
17. Sorensen, P.W.; Sato, K. Second messenger systems mediating sex pheromone and amino acid sensitivity in goldfish olfactory receptor neurons. *Chem. Senses* **2005**, *30* (Suppl. 1), 315–316. [CrossRef]
18. Hansen, A.; Rolen, S.H.; Anderson, K.; Morita, Y.; Caprio, J.; Finger, T.E. Correlation between olfactory receptor cell type and function in the channel catfish. *J. Neurosci.* **2003**, *23*, 9328–9339. [CrossRef]
19. Godfrey, P.A.; Malnic, B.; Buck, L.B. The mouse olfactory receptor gene family. *Proc. Natl. Acad. Sci. USA* **2004**, *101*, 2156–2161. [CrossRef]
20. Rubenstein, D.R.; Lovette, I.J. Reproductive skew and selection on female ornamentation in social species. *Nature* **2009**, *462*, 786–789. [CrossRef]
21. Kurtovic, A.; Widmer, A.; Dickson, B.J. A single class of olfactory neurons mediates behavioural responses to a Drosophila sex pheromone. *Nature* **2007**, *446*, 542–546. [CrossRef] [PubMed]
22. Wyart, C.; Webster, W.W.; Chen, J.H.; Wilson, S.R.; McClary, A.; Khan, R.M.; Sobel, N. Smelling a single component of male sweat alters levels of cortisol in women. *J. Neurosci.* **2007**, *27*, 1261–1265. [CrossRef] [PubMed]
23. Haga, S.; Hattori, T.; Sato, T.; Sato, K.; Matsuda, S.; Kobayakawa, R.; Sakano, H.; Yoshihara, Y.; Kikusui, T.; Touhara, K. The male mouse pheromone ESP1 enhances female sexual receptive behaviour through a specific vomeronasal receptor. *Nature* **2010**, *466*, 118–122. [CrossRef]
24. Roberts, S.A.; Simpson, D.M.; Armstrong, S.D.; Davidson, A.J.; Robertson, D.H.; McLean, L.; Beynon, R.J.; Hurst, J.L. Darcin: A male pheromone that stimulates female memory and sexual attraction to an individual male's odour. *BMC Biol.* **2010**, *8*. [CrossRef] [PubMed]
25. Salvemini, M.; Polito, C.; Saccone, G. Fruitless alternative splicing and sex behaviour in insects: An ancient and unforgettable love story? *J. Genet.* **2010**, *89*, 287–299. [CrossRef] [PubMed]
26. Ishii, K.K.; Osakada, T.; Mori, H.; Miyasaka, N.; Yoshihara, Y.; Miyamichi, K.; Touhara, K. A Labeled-Line Neural Circuit for Pheromone-Mediated Sexual Behaviors in Mice. *Neuron* **2017**, *95*, 123–137. [CrossRef]
27. Yang, L.; Wang, Y.; Zhang, Z.; He, S. Comprehensive transcriptome analysis reveals accelerated genic evolution in a Tibet fish, Gymnodiptychus pachycheilus. *Genome Biol. Evol.* **2015**, *7*, 251–261. [CrossRef]
28. Yang, L.; Jiang, H.; Wang, Y.; Lei, Y.; Chen, J.; Sun, N.; Lv, W.; Wang, C.; Near, T.J.; He, S. Expansion of vomeronasal receptor genes (OlfC) in the evolution of fright reaction in Ostariophysan fishes. *Commun. Biol.* **2019**, *2*, 235. [CrossRef]
29. Yang, L.; Jiang, H.; Chen, J.; Lei, Y.; Sun, N.; Lv, W.; Near, T.J.; He, S. Comparative Genomics Reveals Accelerated Evolution of Fright Reaction Genes in Ostariophysan Fishes. *Front. Genet.* **2019**, *10*, 1283. [CrossRef]
30. Cunningham, F.; Achuthan, P.; Akanni, W.; Allen, J.; Amode, M.R.; Armean, I.M.; Bennett, R.; Bhai, J.; Billis, K.; Boddu, S.; et al. Ensembl 2019. *Nucleic Acids Res.* **2019**, *47*, D745–D751. [CrossRef]
31. Langmead, B.; Salzberg, S.L. Fast gapped-read alignment with Bowtie 2. *Nat. Methods* **2012**, *9*, 357–359. [CrossRef]
32. Li, B.; Dewey, C.N. RSEM: Accurate transcript quantification from RNA-Seq data with or without a reference genome. *BMC Bioinform.* **2011**, *12*, 323. [CrossRef]
33. Robinson, M.D.; McCarthy, D.J.; Smyth, G.K. edgeR: A Bioconductor package for differential expression analysis of digital gene expression data. *Bioinformatics* **2010**, *26*, 139–140. [CrossRef] [PubMed]
34. Eden, E.; Navon, R.; Steinfeld, I.; Lipson, D.; Yakhini, Z. GOrilla: A tool for discovery and visualization of enriched GO terms in ranked gene lists. *BMC Bioinform.* **2009**, *10*, 48. [CrossRef] [PubMed]
35. Barbazuk, W.B.; Fu, Y.; McGinnis, K.M. Genome-wide analyses of alternative splicing in plants: Opportunities and challenges. *Genome Res.* **2008**, *18*, 1381–1392. [CrossRef] [PubMed]
36. Shen, S.; Park, J.W.; Lu, Z.X.; Lin, L.; Henry, M.D.; Wu, Y.N.; Zhou, Q.; Xing, Y. rMATS: Robust and flexible detection of differential alternative splicing from replicate RNA-Seq data. *Proc. Natl. Acad. Sci. USA* **2014**, *111*, E5593–E5601. [CrossRef]
37. Edgar, R.C. MUSCLE: Multiple sequence alignment with high accuracy and high throughput. *Nucleic Acids Res.* **2004**, *32*, 1792–1797. [CrossRef]
38. Stamatakis, A. RAxML version 8: A tool for phylogenetic analysis and post-analysis of large phylogenies. *Bioinformatics* **2014**, *30*, 1312–1313. [CrossRef]

39. Livak, K.J.; Schmittgen, T.D. Analysis of relative gene expression data using real-time quantitative PCR and the 2(T)(-Delta Delta C) method. *Methods* **2001**, *25*, 402–408. [CrossRef]
40. Ibarra-Soria, X.; Levitin, M.O.; Saraiva, L.R.; Logan, D.W. The olfactory transcriptomes of mice. *PLoS Genet.* **2014**, *10*, e1004593. [CrossRef]
41. Chen, M.; Martins, R.N.; Lardelli, M. Complex splicing and neural expression of duplicated tau genes in zebrafish embryos. *J. Alzheimer's Dis. JAD* **2009**, *18*, 305–317. [CrossRef] [PubMed]
42. Seo, M.; Caetano-Anolles, K.; Rodriguez-Zas, S.; Ka, S.; Jeong, J.Y.; Park, S.; Kim, M.J.; Nho, W.G.; Cho, S.; Kim, H.; et al. Comprehensive identification of sexually dimorphic genes in diverse cattle tissues using RNA-seq. *BMC Genom.* **2016**, *17*. [CrossRef] [PubMed]
43. Dang, P.; Fisher, S.A.; Stefanik, D.J.; Kim, J.; Raper, J.A. Coordination of olfactory receptor choice with guidance receptor expression and function in olfactory sensory neurons. *PLoS Genet.* **2018**, *14*, e1007164. [CrossRef] [PubMed]
44. Saraiva, L.R.; Ibarra-Soria, X.; Khan, M.; Omura, M.; Scialdone, A.; Mombaerts, P.; Marioni, J.C.; Logan, D.W. Hierarchical deconstruction of mouse olfactory sensory neurons: From whole mucosa to single-cell RNA-seq. *Sci. Rep.* **2015**, *5*, 18178. [CrossRef] [PubMed]
45. Malnic, B.; Godfrey, P.A.; Buck, L.B. The human olfactory receptor gene family. *Proc. Natl. Acad. Sci. USA* **2004**, *101*, 2584–2589. [CrossRef] [PubMed]
46. Shiao, M.S.; Chang, A.Y.F.; Liao, B.Y.; Ching, Y.H.; Lu, M.Y.J.; Chen, S.M.; Li, W.H. Transcriptomes of Mouse Olfactory Epithelium Reveal Sexual Differences in Odorant Detection. *Genome Biol. Evol.* **2012**, *4*, 703–712. [CrossRef]
47. Olender, T.; Keydar, I.; Pinto, J.M.; Tatarskyy, P.; Alkelai, A.; Chien, M.S.; Fishilevich, S.; Restrepo, D.; Matsunami, H.; Gilad, Y.; et al. The human olfactory transcriptome. *BMC Genom.* **2016**, *17*. [CrossRef]
48. Saraiva, L.R.; Riveros-McKay, F.; Mezzavilla, M.; Abou-Moussa, E.H.; Arayata, C.J.; Makhlouf, M.; Trimmer, C.; Ibarra-Soria, X.; Khan, M.; Van Gerven, L.; et al. A transcriptomic atlas of mammalian olfactory mucosae reveals an evolutionary influence on food odor detection in humans. *Sci. Adv.* **2019**, *5*, eaax0396. [CrossRef]
49. Rodriguez, I.; Greer, C.A.; Mok, M.Y.; Mombaerts, P. A putative pheromone receptor gene expressed in human olfactory mucosa. *Nat. Genet.* **2000**, *26*, 18–19. [CrossRef]
50. Zhang, X.; Rogers, M.; Tian, H.; Zou, D.J.; Liu, J.; Ma, M.; Shepherd, G.M.; Firestein, S.J. High-throughput microarray detection of olfactory receptor gene expression in the mouse. *Proc. Natl. Acad. Sci. USA* **2004**, *101*, 14168–14173. [CrossRef]
51. Alioto, T.S.; Ngai, J. The odorant receptor repertoire of teleost fish. *BMC Genom.* **2005**, *6*, 173. [CrossRef] [PubMed]
52. Cao, Y.; Oh, B.C.; Stryer, L. Cloning and localization of two multigene receptor families in goldfish olfactory epithelium. *Proc. Natl. Acad. Sci. USA* **1998**, *95*, 11987–11992. [CrossRef] [PubMed]
53. Saraiva, L.R.; Korsching, S.I. A novel olfactory receptor gene family in teleost fish. *Genome Res.* **2007**, *17*, 1448–1457. [CrossRef]
54. Holman, L.; Helantera, H.; Trontti, K.; Mikheyev, A.S. Comparative transcriptomics of social insect queen pheromones. *Nat. Commun.* **2019**, *10*, 1593. [CrossRef] [PubMed]
55. McKenzie, S.K.; Kronauer, D.J.C. The genomic architecture and molecular evolution of ant odorant receptors. *Genome Res.* **2018**, *28*, 1757–1765. [CrossRef] [PubMed]
56. McKenzie, S.K.; Fetter-Pruneda, I.; Ruta, V.; Kronauer, D.J. Transcriptomics and neuroanatomy of the clonal raider ant implicate an expanded clade of odorant receptors in chemical communication. *Proc. Natl. Acad. Sci. USA* **2016**, *113*, 14091–14096. [CrossRef]
57. Zhou, S.; Stone, E.A.; Mackay, T.F.; Anholt, R.R. Plasticity of the chemoreceptor repertoire in Drosophila melanogaster. *PLoS Genet.* **2009**, *5*, e1000681. [CrossRef]
58. Clowney, E.J.; Magklara, A.; Colquitt, B.M.; Pathak, N.; Lane, R.P.; Lomvardas, S. High-throughput mapping of the promoters of the mouse olfactory receptor genes reveals a new type of mammalian promoter and provides insight into olfactory receptor gene regulation. *Genome Res.* **2011**, *21*, 1249–1259. [CrossRef]
59. Magklara, A.; Yen, A.; Colquitt, B.M.; Clowney, E.J.; Allen, W.; Markenscoff-Papadimitriou, E.; Evans, Z.A.; Kheradpour, P.; Mountoufaris, G.; Carey, C.; et al. An epigenetic signature for monoallelic olfactory receptor expression. *Cell* **2011**, *145*, 555–570. [CrossRef]
60. Stowers, L.; Logan, D.W. Sexual dimorphism in olfactory signaling. *Curr. Opin. Neurobiol.* **2010**, *20*, 770–775. [CrossRef]

61. Miklosi, A.; Andrew, R.J. The Zebrafish as a Model for Behavioral Studies. *Zebrafish* **2006**, *3*, 227–234. [CrossRef] [PubMed]
62. Gerlai, R. Zebra fish: An uncharted behavior genetic model. *Behav. Genet.* **2003**, *33*, 461–468. [CrossRef] [PubMed]

Review

The Role of Olfactory Genes in the Expression of Rodent Paternal Care Behavior

Tasmin L. Rymer [1,2,3]

[1] College of Science and Engineering, James Cook University, P. O. Box 6811, Cairns, QLD 4870, Australia; tasmin.rymer@jcu.edu.au; Tel.: +61-7-4232-1629

[2] Centre for Tropical Environmental and Sustainability Sciences, James Cook University, P. O. Box 6811, Cairns, QLD 4870, Australia

[3] School of Animal, Plant and Environmental Sciences, University of the Witwatersrand, Private Bag 3, WITS, Johannesburg 2050, South Africa

Received: 30 January 2020; Accepted: 5 March 2020; Published: 10 March 2020

Abstract: Olfaction is the dominant sensory modality in rodents, and is crucial for regulating social behaviors, including parental care. Paternal care is rare in rodents, but can have significant consequences for offspring fitness, suggesting a need to understand the factors that regulate its expression. Pup-related odor cues are critical for the onset and maintenance of paternal care. Here, I consider the role of olfaction in the expression of paternal care in rodents. The medial preoptic area shares neural projections with the olfactory and accessory olfactory bulbs, which are responsible for the interpretation of olfactory cues detected by the main olfactory and vomeronasal systems. The olfactory, trace amine, membrane-spanning 4-pass A, vomeronasal 1, vomeronasal 2 and formyl peptide receptors are all involved in olfactory detection. I highlight the roles that 10 olfactory genes play in the expression of direct paternal care behaviors, acknowledging that this list is not exhaustive. Many of these genes modulate parental aggression towards intruders, and facilitate the recognition and discrimination of pups in general. Much of our understanding comes from studies on non-naturally paternal laboratory rodents. Future studies should explore what role these genes play in the regulation and expression of paternal care in naturally biparental species.

Keywords: discrimination; main olfactory system; olfaction; paternal care; recognition; vomeronasal system

---

1. Introduction

In order to survive, all animals must detect, interpret and respond to an array of sensory information in their immediate environment [1]. Animals need to locate and assess the quality of food, detect and avoid predators, and identify mates and competitors [1]. Animals can gain information about these resources or threats, as well as convey information to other individuals, via numerous modalities, including vision and olfaction. For many mammals, particularly rodents, olfaction is most likely the dominant sensory modality [2,3]. Social behaviors, including parent/offspring interactions, are also strongly regulated by olfactory cues (e.g., prairie voles (*Microtus ochrogaster*) [4] and Syrian hamsters (*Mesocricetus auratus*) [5]).

While mammalian maternal care is essential for offspring survival, mammalian paternal care is rare (5%–10% of species [6]). This is most likely because the costs associated with paternal care (e.g., predation risk [7], increased energetic expenditure [8], loss of mating opportunities [9], and reduced survival [10]), and the inability of males to physically associate with offspring during prenatal development [11], are major limiting factors in the evolution of paternal care [12]. However, paternal males can significantly influence offspring growth, survival, and cognitive and behavioral development [6]. Consequently, it is necessary to understand what factors play a role in the expression of paternal care when it does occur.

The ability to detect, recognize, and discriminate olfactory cues of social significance between fathers and offspring is dependent on complex neural mechanisms that may be regulated by specific olfactory genes. Here, I consider the role of olfaction in the expression of paternal care behavior in rodents. I first describe the behavioral machinery [13] of paternal care in general, and how this is connected with brain regions associated with the detection and interpretation of olfactory cues. I then discuss the genetic regulation of olfaction in the different olfactory systems in general. Finally, I discuss how olfactory cues might regulate the expression of paternal care behavior, highlighting the roles that 10 different olfactory genes may play in the expression of paternal care behavior. This list is not exhaustive, and there are likely multiple other genes that could be equally important. While indirect paternal care (e.g., alarm calling, [14]) is an important component of the paternal repertoire, I only focus on direct paternal care behavior (retrieval, huddling, nest building and grooming) because indirect paternal care does not require direct pup contact, whereas direct paternal care behavior does. I rely extensively on studies from laboratory mice, which are not naturally paternal, because the predominant literature on olfactory regulation of paternal care behavior comes from these studies, and the literature is decidedly depauperate on how the genetic mechanisms of olfaction moderate paternal care in biparental species. Nevertheless, these studies provide a starting point for those interested in mechanisms underlying the expression of paternal care in biparental species.

2. Neural Regulation of Paternal Care Behavior

2.1. Brain Regions Implicated in the Regulation and Expression of Paternal Care

The most important brain region associated with paternal care behavior is the hypothalamic medial preoptic area (MPOA [15,16]). The MPOA is anatomically connected to the bed nucleus of the stria terminalis (BNST) and the amygdala [17,18], as well as the lateral preoptic area (LPOA [19]), and the adjoining substantia innominata (SI [20]). Lateral efferent neurons project from the MPOA to the LPOA and SI, and pass through the lateral hypothalamus (LH) to the ventral tegmental area (VTA [19]). Neuronal disruption to the central MPOA and the lateral efferent neurons [21,22] can disrupt paternal care behavior, specifically retrieval behavior. Different subregions or neuron populations in the MPOA may affect an individual's responsiveness to pup-specific odor cues [23].

The amygdala may also play an important role in the regulation of paternal care because of its distinctive neuronal heterogeneity, specifically, the caudal olfactory cortex [24]. The olfactory tubercle receives direct information from the main olfactory bulb (OB), whereas the medial amygdalar nucleus (meA), which lies adjacent to the olfactory tubercle, receives direct information from the accessory olfactory bulb (AOB [25]).

2.2. Olfactory Systems and Associations with Brain Regions Implicated in Paternal Care

Mammalian olfactory systems are complex, remarkably precise (one odorant receptor gene expressed per cell; [26]) and allow mammals to recognize and discriminate a large diversity of odorant molecules [27]. There are two main, anatomically and functionally distinct chemoreceptor systems [28], namely the main olfactory system (MOS) and the vomeronasal system (VNS). It has been suggested that the MOS primarily detects volatile odorants from the environment [29,30], whereas the VNS primarily detects non-volatile odorants from conspecifics [28,30,31], although both systems can, to a degree, detect both types of odor cues [30]. The Grueneberg ganglion, a chemosensory organ that appears to mediate behavioral responses to alarm pheromones in rodents [32], and the septal organ of Masera, a patch of sensory epithelium separate to the MOE that may have a dual role in surveying food or conspecific sexual odors [33], are not considered here.

2.2.1. The Main Olfactory System (MOS)

The nose houses the MOS, which consists of the main olfactory epithelium (MOE; Figure 1). This is the primary site for the detection of volatile odorants [1]. The main olfactory sensory neurons (OSNs) in the MOE, of which there are approximately 10 million in vertebrates, are located directly in the nasal airstream; thus, stimulus access simply requires passive respiration or a sniffing action [34]. The 500–1000 olfactory receptors (ORs [1,27,35]) that belong to the rhodopsin-like G-protein-coupled receptor superfamily (GPCRs [26,35,36]), are located in the cell membranes of the OSNs, and bind specific odorant ligands [35]. While the ORs are responsible for detecting chemosensory cues, they are also involved in axonal guidance to the brain [36,37].

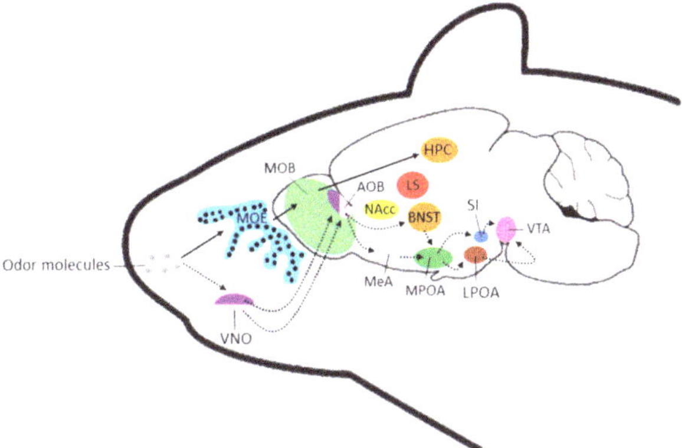

Figure 1. Schematic representation of transmission of olfactory information via the main olfactory system (solid black arrows) or the vomeronasal system (broken arrows) to corresponding brain regions. Black circles in the MOE indicate the broad localization of olfactory sensory neurons. Note: locations of brain regions not exact and for pictorial purposes only. AOB: accessory olfactory bulb; BNST: bed nucleus of the stria terminalis; HPC: hippocampus; LS: lateral septrum; MOB: main olfactory bulb; MOE: main olfactory epithelium; MeA: medial amydala; LPOA: lateral preoptic area; MPOA: medial preoptic area; NAcc: nucleus accumbens; SI: substantia innominata; VNO: vomeronasal organ; VTA: ventral tegmental area.

Neurons that express the same OR type converge at similar sites (glomeruli) within the OB [38], forming synapses with mitral [36,38] or tufted cells [39] and conveying olfactory information from the MOE to the OB [40]. The signals from each OR are transferred to the anterior olfactory nucleus, the cortical amygdala [30] and a small number of pyramidal cells that form clusters in the piriform cortex (PC) of the olfactory cortex [41]. The PC cells project to the orbitofrontal cortex, amygdaloid cortex, prefrontal cortex, perirhinal cortex and entorhinal cortex, through which they access the hippocampus (Figure 1; [42]). Androgen and estrogen cellular receptors are expressed in the PC [43], suggesting a responsiveness to hormones that may regulate sexual and paternal behaviors [3]. Importantly, olfactory bulbectomy negatively affects paternal care in male prairie voles [44].

2.2.2. The Vomeronasal System (VNS)

Closely associated with the MPOA is the accessory olfactory or VNS (Figure 1; [45]). The VNS is primarily involved with the reception and decoding of olfactory cues, providing a relatively direct pathway to the amygdala [46], BNST [39] and hypothalamic areas [47].

The vomeronasal organ (VNO; Figure 1) of the VNS is an extraordinarily sensitive structure [48]. In rodents, the morphological complexity of the VNO is greater than any other mammal [49]. The VNO detects both volatile and non-volatile olfactory signals [2], and neurons that express the same receptor form multiple glomeruli within the AOB [34]. Neural projections transfer the olfactory signals to several brain regions via the AOB [50], including the corticomedial amygdala [24] and the BNST (Figure 1; [51]). Projections from the AOB also extend to the medial (Figure 1) and posteromedial cortical (C3) amygdaloid nuclei [51] and the ventral hypothalamus [52]. Interestingly, there is sexual dimorphism in the AOB [53], meA [54] and BNST [54,55].

The MPOA then receives these impulses (Figure 1), activating Galanin-expressing neurons ($MPOA^{Gal}$ [22]), and this cascade of impulses then activates neurons in the LPOA and SI (Figure 1; [20]). From here, LPOA descending efferent neurons, which are localized in the dorsal LH [19], trigger neurons in the VTA (Figure 1). The VTA is part of the dopaminergic reward system, and is associated with reinforcement learning, with unweaned offspring being a strong reinforcing stimulus to males [56]. This pathway likely influences the processing of pup-related olfactory cues, and mediates and regulates pup-directed aggression [57]. Disruptions to this pathway inhibit infanticide and promote paternal care [22,58].

3. Genetic Regulation of Olfaction

3.1. Genetic Regulation in the Main Olfactory System

Mammals can recognize and discriminate thousands of odor molecules due to a large multigene family (±1400 functional genes [29]) in the MOE that encodes the ORs. Each OSN expresses only one or a few odorant receptor genes [2,36,59,60], and the genes are randomly monoallelically expressed (i.e., half express the maternal allele while the other half express the paternal allele [36,59]). The ORs are bound to the G-protein $G_{\alpha olf}$ [30] and are typically not very selective. Thus, an OSN typically responds to a range of related odor cues (i.e., combinatorial in nature [61]). In mammals, ORs fall into two major groups (phylogenetic clades; [62]): "Class I" ORs (known as fish-like receptors as they were first identified in fish) comprise approximately 10% of functional ORs, while "Class II" ORs (mammalian-like receptors) comprise approximately 90% [40]. It is thought that a subset of some of these ORs from both classes, respond to volatile compounds in food, thereby influencing foraging behavior and food preferences [63].

In addition to the ORs, trace amine receptors (TAARs) are also expressed by neurons localized in the MOE [63–65], and are activated by distinct combinations of volatile amines, many of which occur in urine [66]. There are between 15 and 17 TAARs found in rodents [64,66]. TAARs are expressed in a small number of OSNs [66] and, like ORs, they are expressed in a mutually exclusive manner [64]. Interestingly, TAARs in the MOE are localized to $G_{\alpha olf}$–expressing sensory neurons that can stimulate cyclic adenosine monophosphate (cAMP) pathways, indicating that they couple to canonical olfactory pathways [66]. In rodents, $G_{\alpha olf}$ is highly expressed in the medium spiny neurons of the striatum, which houses the dopamine 1 receptor [67] and is critical for transduction of the ORs and complete olfactory function [68].

Lastly, a set of molecularly atypical neurons residing in the MOE expresses other non-GPCR receptors (guanylate cyclase GC-D) that are encoded by membrane-spanning 4-pass A (Ms4a) genes [63]. While every Ms4a protein detects specific odors [69], each is likely to play a role in regulating the social acquisition of food preference via olfactory cues [70].

3.2. Genetic Regulation in the Vomeronasal System

In the VNS, the VNS sensory neurons (VSNs) are located away from the nasal airstream, and activation of the neurons thus requires a vascular pumping mechanism [71]. This mechanism enables the VNO to take up non-volatile stimuli that are investigated by direct nasal contact [34]. The VSNs are among the most sensitive of mammalian chemoreceptors [72]. In contrast to the main olfactory genes, VNO receptors detect only a limited group of ligands (differential tuning hypothesis [73]). There are 250–300 functional vomeronasal receptor genes [74], in at least three families, including vomeronasal type 1 receptor genes (V1Rs), vomeronasal type 2 receptor genes (V2Rs), and formyl peptide receptor genes (FPRs). As for the ORs in the MOE, the V1Rs and V2Rs encode G- protein-coupled transmembrane proteins [28,39,74].

The main receptor proteins of the VNO consist of the V1R (±150) and V2R (±160) families of vomeronasal receptors [75], each derived from individual genes [29]. The two families of receptors are expressed in anatomically distinct neuronal populations of the VN epithelium [2] that coincide with different zones of G-protein expression (V1R = apical zone of the epithelium, express $G_{\alpha i2}$, dark purple in Figure 1; [28]; V2R = basal zone of the epithelium, express $G_{0\alpha}$, (light purple in Figure 1; [39]), and each VNO neuron expresses only a single receptor protein [75]. In addition, the two classes project to anatomically and functionally separate sub-regions of the AOB, suggesting differential processing of vomeronasal stimuli [76]. V1R-expressing neurons project to the anterior sub-region of the AOB, while V2R-expressing neurons project to the posterior sub-region of the AOB [2]. However, neural projections coming from each of these two regions then project to and overlap at the level of the amygdala, the accessory olfactory tract and the BNST [77].

Unlike OSNs, the VSNs are highly selective for individual molecules [2], although due to their highly diverse nature, V1Rs typically respond to a wide variety of different odor molecules [75,78], and are known to respond to the urinary volatiles 2,3-dehydro-exo-brevicomin (DB) and 2-sec-butyl-4,5-dihydrothiazole (BT, [72]). Both V1Rs and V2Rs are thought to detect olfactory cues that are related to conspecifics [28,31]. For example, 129/SvEv male mice with a cluster of V1R genes genetically deleted show reduced sexual behavior [79]. Interestingly, H2-Mv (a class of major histocompatibility complex (MHC) proteins) is coexpressed in V2Rs in rodents [80], with M10 and M1 family proteins being expressed exclusively in the V2Rs [81]. Indeed, it has been suggested that correct V2R expression relies on the M10s [81]. H2-Mv genes are not randomly expressed, and certain combinations of genes are located with particular V2Rs [82], which could explain an individual's responsiveness to particular MHC-associated chemosignals [2].

The FPRs are another family of olfactory neurons expressed by localized VNO neurons [63,83]. Interestingly, FPR olfactory expression is restricted to rodents [84], and expression of these receptors occurs in a punctate and monogenic pattern in the VSNs [83], which is characteristic of the transcription of olfactory chemoreceptor genes [85]. Within the vomeronasal (VN) epithelium, Fpr-rs3, -rs4, -rs6 and –rs7 are transcribed by neurons in the apical zone, coincident with V1Rs, while Fpr-rs1 is transcribed by neurons in the basal zone, coincident with V2Rs [83]. It has been suggested that FPRs play a role in the detection of pathogens or pathogenic states [83].

4. Olfaction and Paternal Care Behavior: Suggested Genetic Regulation

Numerous candidate genes, many coding for hormone expression, influence paternal care behaviors (e.g., estrogen receptor alpha (ERα) [86]). However, the regulation of paternal care is likely under multisensory control, and olfactory stimuli from pups should be neurally integrated to allow males to recognize offspring [23] and provide paternal care accordingly. The use of odor for distinguishing kin relationships [87,88] and for paternal kin discrimination (e.g., golden hamsters (*Mesocricetus auratus*) [89]) is well documented in rodents. For example, disruption or damage to the OB diminishes paternal care behaviors in biparental male prairie voles [15]. However, several olfactory genes, or genes that regulate olfactory processes, could be involved in the regulation of paternal care behaviors (Table 1).

The MOS likely has less of an influence on the expression of paternal care behaviors since MOE receptor gene sequences are conserved across both paternal and non-paternal vertebrate species [30]. However, VNO receptors detect a limited group of ligands [73], and there is species-specific variation in VNO receptor diversity [30], suggesting that the differential detection and signaling in the VNO could be important for the expression of paternal care behavior in biparental species compared to non-paternal species. However, it is equally plausible that the MOS and VNS work synergistically in identifying, recognizing, and discriminating pup odor cues, and that paternal care is mediated by both systems. Below, I discuss 10 genes that likely work mutually to regulate the expression of paternal care in male rodents.

Table 1. Genes involved in the regulation and expression of paternal care behaviors in rodents, their theorized olfactory location, functioning associated brain regions and interactions with other genes, proteins or hormones (all references provided in text).

Gene	Location	Associated Brain Regions	Interactions	Effect on Behavior
$G_{\alpha i2}$	V1Rs in the apical zone of the VN epithelium	AOB, MPOA, amygdala, BNST, hypothalamus, LPOA, SI, LH, VTA	*Trp2*, FPRs and ORs; *c-Fos*, *fosB*, *CREB*, *ERK*. *SPRY1*, *Rad*, MUPs, *ESP1*	Genetic KO = ↓ aggression, ↑ grooming, ↑ retrieval
Trp2	V1Rs and V2Rs in the VNO	AOB, MPOA, amygdala, BNST, hypothalamus, LPOA, SI, LH, VTA	$G_{\alpha i2}$, Gal^+, *c-Fos*, *fosB*, *CREB*, *ERK*. *SPRY1*, *Rad*, $Esr1^+$	Genetic KO = ↓ aggression, ↓ recognition, ↓ nest-building, ↓ time with pups
CD38	ORs in the MOS	MPOA, BNST, LS, amygdala, AOB, OB, SON, NAcc, olfactory nucleus, PC, orbitofrontal, prefrontal, perirhinal cortex and entorhinal cortices, hippocampus	OT, GABA, *c-Fos*, calcium ion signaling molecules, cyclic ADP-ribose	Genetic KO = ↓ retrieval, ↓ grooming, ↓ huddling
Olfr692	ORs in the VNO	AOB, MPOA, amygdala, BNST, hypothalamus, LPOA, SI, LH, VTA	*Egr1*, $G_{\alpha o+}$	↑ expression (virgin ♂s) = ↑ aggression, ? paternal care
MUP genes	TAARs in the MOE; V2Rs in the VNO	OB, AOB, MPOA, amygdala, BNST, hypothalamus, LPOA, SI, LH, VTA, LS, SON, NAcc, olfactory nucleus, PC, orbitofrontal, prefrontal, perirhinal and entorhinal cortices, hippocampus	$G_{\alpha i2}$	↑ kin discrimination, ↑ aggression to non-kin
c-Fos	ORs in the MOE; VRs in the VNO	OB, AOB, PC, MPOA, amygdala, BNST, hypothalamus, LPOA, SI, LH, VTA, LS, SON, NAcc, olfactory nucleus, orbitofrontal, prefrontal, perirhinal and entorhinal cortices, hippocampus	*fosB*, *CREB*, $G_{\alpha i2}$, BDNF, *CD38*, *Trp2*, ERK, *SPRY1*, *Rad*, $G_{\alpha olf}$, *Adyc3*, CNG, $Esr1^+$, Gal^+, $ER\alpha$	↑ paternal care
fosB	ORs in the MOE; VRs in the VNO	OB, AOB, PC, MPOA, amygdala, BNST, hypothalamus, LPOA, SI, LH, VTA, LS, SON, NAcc, olfactory nucleus, orbitofrontal, prefrontal, perirhinal and entorhinal cortices, hippocampus	*c-Fos*, *CREB*, $G_{\alpha i2}$, BDNF, *CD38*, *Trp2*, ERK, *SPRY1*, *Rad*, $G_{\alpha olf}$, *Adyc3*, CNG, $Esr1^+$, Gal^+, $ER\alpha$	Genetic KO = ↓ retrieval
CREB	ORs in the MOE; VRs in the VNO	OB, AOB, PC, MPOA, amygdala, BNST, hypothalamus, LPOA, SI, LH, VTA, LS, SON, NAcc, olfactory nucleus, orbitofrontal, prefrontal, perirhinal and entorhinal cortices, hippocampus	*c-Fos*, *fosB*, $G_{\alpha i2}$, BDNF, *CD38*, *Trp2*, ERK, *SPRY1*, *Rad*, $G_{\alpha olf}$, *Adyc3*, CNG, $Esr1^+$, Gal^+, $ER\alpha$	↑ paternal care
Adyc3	ORs in the MOE	Amygdala, MPOA, BNST, LS, AOB, OB, SON, NAcc, olfactory nucleus, PC, orbitofrontal, prefrontal, perirhinal and entorhinal cortices, hippocampus	*c-Fos*, *fosB*, *CREB*, dopamine, $G_{\alpha olf}$	Genetic KO = ↓ general pup recognition
PRLR	ORs in the MOE; V1Rs in the apical epithelium	Choroid plexus, SVZ, dentate gyrus, OB, hippocampus, amygdala, MPOA, BNST, LS, AOB, OB, SON, NAcc, olfactory nucleus, PC, orbitofrontal, prefrontal, perirhinal and entorhinal cortices	Dopamine, AVP	↑ paternal care, ↑ kin pup recognition

V1R: vomeronasal type 1 receptor; V2R: vomeronasal type 2 receptor; OB: main olfactory bulb; AOB: accessory olfactory bulb; MPOA: medial preoptic area; BNST: bed nucleus of the stria terminalis; LPOA: lateral preoptic area; SI: substantia innominate; LH: lateral hypothalamus; KO: knockout; VNO: vomeronasal organ; MOS: main olfactory system; OR: olfactory receptors; OT: oxytocin; MOE: main olfactory epithelium; VTA: ventral tegmental area; SON: supraoptic nucleus; SVZ: subventricular zone; NAcc: nucleus accumbens; FPR: formyl peptide receptor genes; CNG: cyclic nucleotide-gated; BDNF: brain-derived neurotrophic factor; AVP: arginine vasopressin.

4.1. $G_{\alpha i2}$

VSNs in the apical layer of epithelium in the VNS express $G_{\alpha i2}$ (Table 1), and these cells have been implicated in moderating pup-directed aggression by detecting pup odor cues, major urinary proteins (MUPs), and odor cues from the facial area, such as exocrine gland-secreted peptide 1 (*ESP1*; Table 1; [90–92]). Pup odors activate regions of the OB and AOB (Table 1) that are innervated by $G_{\alpha i2}$ neurons [58], prompting aggression. However, male mice with deletion of the $G_{\alpha i2}$ gene were less aggressive towards pups, and showed increased grooming and retrieval of pups (Table 1), most likely because activation in the MPOA was increased in $G_{\alpha i2}{}^{-/-}$ males [93]. It is likely that $G_{\alpha i2}$ activates *Trp2* and calcium ion entry downstream of V1R activation (Table 1), moderating aggression [93]. *Gαi2* may also interact with FPRs and ORs in the VNO [83,94] to mediate pup-directed aggression and paternal care (Table 1).

4.2. *Trp2 (or Trpc2)*

In the VNO, activation of vomeronasal GPCRs causes a phospholipase C-dependent cascade [81], which regulates the *Trp2* cation channel [81,95]. Consequently, *Trp2* plays an important role in signal transduction in both V1R and V2R-expressing neurons [75,95]. Furthermore, *Trp2* plays a role in the expression of aggression in males (Table 1), with male *Trp2*$^{-/-}$ mice showing deficiency in social recognition of conspecifics [81], and a reduction in aggression in resident-intruder style tests [95,96]. These studies indicate that aggression requires a functional VNO. While commonly associated with sexual behaviors, *Trp2* could mediate paternal care by reducing aggression in males that might otherwise be directed towards their own pups. Inactivation of the *Trp2* channel does not impair detection of MHC peptides [30], most likely because MCH genes are expressed in a subpopulation of basilar VNS neurons, with M10 in particular being related to the expression of V2Rs [81]. This suggests that males could still identify their own pups via an MHC signature, as sensory neurons respond to MHC peptides in both the VNO and the MOE [97]. If *Trp2* in biparental males is deactivated while cohabiting with a female during the gestation period, or by olfactory cues from the pups themselves, this could cause males to respond paternally rather than aggressively towards their young [22]. Female *Trp2*$^{-/-}$ mice show impaired nest building behavior and time spent with pups [98,99], further suggesting that *Trp2* could also be important for regulating some direct paternal care behaviors as well (Table 1).

4.3. *CD38*

Another gene that may play a role in the expression of paternal care behavior is *CD38*, a transmembrane glycoprotein that catalyzes the formation of calcium ion signaling molecules [100] and cyclic ADP-ribose (Table 1; [101]). *CD38*$^{-/-}$ mice do not show deficits in olfactory-guided foraging or habituation to non-social stimuli, indicating that genetic knockout of this gene does not impair olfactory function per se [100]. *CD38* is implicated in the release of the neuropeptide oxytocin (OT; (Table 1) from hypothalamic neurons [100,102]. OT is involved in sexual, affiliative and parental care behaviors [102,103], and can stimulate the release of prolactin, in concert with arginine vasopressin (AVP [104,105]. Neurons in the MPOA can activate dopaminergic neurons in the VTA, which innervate GABA neurons in the nucleus accumbens (NAcc; (Table 1). OT receptors are also found in the NAcc, and male *CD38*$^{-/-}$ mice show reduced expression of OT in the NAcc [101]. The posterior pituitary secretes OT into the general circulation [100], and increased OT receptor binding in the BNST, lateral septum (LS), lateral amygdala and accessory olfactory nucleus is associated with increased paternal care in male meadow voles (*Microtus pennsylvanicus*) (Table 1; [106]).

Exposure to offspring olfactory cues activates the mitral cells of the OB [107], and increases OT expression in the supraoptic nucleus (SON) of male mandarin voles *Lasiopodomys mandarinus* [108] and in the MPOA of male California mice *Peromyscus californicus* (Table 1; [109]). *CD38*$^{-/-}$ fathers show consistently decreased levels of plasma and cerebrospinal OT, and concomitantly a reduction in paternal care (Table 1; [100]). *CD38*$^{-/-}$ fathers fail to retrieve pups, and show a reduction in pup grooming, crouching and huddling (Table 1; [101]). Since olfactory function in general is not impaired, genetic knockout of the *CD38* gene indicates an inability to identify odor cues specifically related to pups.

4.4. *Olfr692*

Olfr692 is a member of the OR gene family that is highly expressed in the basal zone of the VNO (Table 1) of adult male mice [1,110], with expression levels similar to those of *Vmn1r188* and *Vmn2r118* in the V1R and V2R families [1]. An extensive number of *Olfr692*-positive cells occurs in the VNOs of adult rodents, but expression is virtually absent in juveniles [1]. This expression pattern contrasts that of the expression pattern of VR genes, which are first expressed in embryos [28], and the few VNO ORs that are mostly expressed in juveniles [94]. This differential expression of *Olfr692* in adults and juveniles suggests that *Olfr692* may play a role in adult-specific behaviors [1].

Olfr692-positive cells in the VNO are activated by odor cues from pups [1]. After exposure to pups, virgin male mice show considerable activation of these cells [1], increasing expression of the immediate early gene *Egr1* (Table 1; [110]). However, activation appears to be dependent on prior social and parenting experience, as males that have sired and cared for pups show low activation of the *Olfr692*-expressing neurons, which could be modulated by endocrine mechanisms [1]. This differential expression of *Olfr692* between fathers and non-fathers suggests that *Olfr692* may mediate aggression and infanticide towards novel pups (Table 1; [22,58]). This could be the neural "switch" that results in infanticidal males becoming paternal [111]. Olfactory cues sensed during active sniffing and investigation of the young [23] could activate or alter the activity of MPOA or BNST neurons (Table 1), leading to this switch in behavior [22,58,112], although the absolute role of *Olfr692* in the expression of paternal care behaviors still requires testing.

4.5. MUP Genes

Urinary volatile pheromones are bound to highly polymorphic MUPs [113], a polymorphic group known to be important in chemosensory communication [114], and which are potentially detected by TAARs in the MOE (Table 1; [64]). MUPs are synthesized in the liver, and rodents produce 4-15 MUP variants [114], which may interact directly with the chemosensory receptors to provide a reliable signal of individuality [115]. There are approximately 35 genes in the MUP gene cluster on chromosome 4 [116], and both males and females produce MUPs, indicating that they function as chemical signals for both sexes [114]. While MUPs appear to be principally involved in scent-marking communication [114], MUPs might also function to deliver small semiochemicals to the VNO or MOE [117], thus they may have a similar role to odorant binding proteins [118]. MUPs also act through V2Rs to control interspecies defensive, and intra-species aggressive, behaviors [63], and may also be used for kin discrimination (Table 1; [119]). *MUP3* and *MUP20* elicit aggression in male mice (Table 1; [92]); however, but if males recognize the odor cues from their own offspring, this could potentially deactivate MUP genes, leading to a reduction in aggression and promotion of paternal care.

4.6. c-Fos, fosB and CREB

c-Fos and *fosB* are immediate early genes (genes that are rapidly expressed in response to a stimulus [120]). The expression of *c-Fos* can be increased by abiotic (e.g., light [121]) or social (e.g., Syrian hamsters [122]) stimuli and, as *fosB* is homologous to *c-Fos*, it follows a similar induction pattern [23]. *fosB* is expressed in several brain regions, including the OB, AOB and PC (Table 1; [123]). Both *c-Fos* and *fosB* are expressed in the MPOA (Table 1; [52,123,124]), particularly in response to pup exposure, which indicates that *c-Fos* and *fosB* neurons are involved in general pup recognition [123,124], regardless of kin relationship. However, the exact neural properties and connections of *c-Fos* positive neurons in the MPOA (Table 1) during active parental care is not well known [23]. Interestingly, while male genetic knockouts for the *fosB* gene showed no impairment in general olfactory discrimination [23], they were less paternal towards pups, and were impaired in their retrieval responses (Table 1; [123]), suggesting that recognition of pups (regardless of kin relationship) specifically is impaired.

Pup-related olfactory stimuli could activate MPOA via activation of brain-derived neurotrophic factor (BDNF), *CD38* or *Trp2* by $G_{\alpha i2}$ (Table 1), stimulating calcium ion channels [124]. Extracellular signal regulated kinase (ERK) is then phosphorylated in the MPOA neurons (Table 1), inducing transcription of *c-Fos* and *fosB*, which could cause the upregulation of *SPRY1* and *Rad*, leading to increasing paternal care (Table 1; [124]). However, it is possible that, in males, ERK also works in concert with the Ca^{2+}/cAMP-responsive element-binding protein (*CREB;* Table 1), as *CREB* increases in the female MPOA following pup exposure [125], and affects maternal behavior [126]. $G_{\alpha olf}$ [127], *Adyc3* [128], and cyclic nucleotide gated (CNG) cation channels [129] are all enriched in the MOE (Table 1), indicating an important role for cAMP in olfactory signaling [130]. Pup exposure stimulates *CREB*, leading to an increased calcium influx in MPOA neurons that express *Esr1*$^{+}$ and *Gal*$^{+}$ [131,132], and *Gal*$^{+}$

neurons are known to regulate paternal care in mice (Table 1; [22]). Alternately, *CREB* may activate *ERα* [125], which could then affect paternal care (Table 1).

4.7. *Adyc3*

The adenylate cyclase type 3 (*Adcy3*) gene encodes type 3 adenylyl cyclase (AC3 [133]), which is coupled to some odorant receptors [130]. Both *Adyc3* and $G_{\alpha olf}$ are required for sensory transduction in the MOE (Table 1; [133]). *Adyc3* is also expressed in several brain regions, including the amygdala and MPOA (Table 1; [130]), suggesting a potential role in parental behavior. Female genetic knockouts for *Adyc3* and $G_{\alpha olf}$ show impaired pup retrieval, nest building and huddling behaviors [68,130]. These responses are likely a consequence of an inability to detect pup odor cues [130]. Similarly, male *Adyc3*$^{-/-}$ males are anosmic, and unable to detect pup odors (Table 1), suggesting that cAMP plays a role in olfactory signaling in males [134]. Interestingly, aggression may also be mediated by *Adyc3*, as female *Adyc3*$^{-/-}$ mice are not aggressive to an intruder that represented a threat to pups, and were not aggressive towards alien pups [130]. It is further possible that motivation to provide paternal care is driven by dopamine through modulation of dopamine type 1 (D1) receptor-dependent cAMP signaling by $G_{\alpha olf}$ (Table 1; [135]).

4.8. *PRLR*

Prolactin is a gonadotropic hormone secreted by the anterior pituitary [136], but inhibited by dopamine from the hypothalamus (Table 1; [137]). It can cross the blood-brain barrier via a receptor-mediated transport mechanism in the choroid plexus [138], entering the cerebrospinal fluid, and exerting a direct influence in the brain (Table 1; [136,137]). Circulating prolactin increases before parturition in some paternal species [139], and might be critical for organizing neural substrates associated with paternal care behaviors [140].

PRLR is an imprinted gene that is expressed at low levels in the OB, and mediates paternal-offspring recognition via olfactory neurogenesis (Table 1; [86]). Mak and Weiss [141] found that neurogenesis under the influence of prolactin signaling increased in in the subventricular zone (SVZ) and dentate gyrus of male house mice (*Mus domesticus*) following interactions with pups (Table 1). Similarly, prolactin receptor mRNA transcript levels increased in the choroid plexus of male Djungarian hamsters (*Phodopus campbelli*) during the early postnatal period [142]. Some of the cells in the mouse SVZ and dentate gyrus matured into olfactory interneurons, and responded preferentially to offspring odors compared to other odor types, indicating a central role of *PRLR* in offspring recognition (Table 1; [141]).

5. Conclusions

Pup-related odor cues are critical for the onset and maintenance of mammalian paternal care. However, there are numerous genetic mechanisms underlying the detection, recognition and discrimination of rodent pups, which suggests complex modulation of paternal care behaviors. In this review, I discussed 10 genes that have been implicated in the regulation of paternal care via pup-related olfactory cues in rodents. There are likely many more. That paternal care is likely under multisensory control further complicates our understanding of the direct effects of olfactory genes on the regulation of paternal care behaviors. Since much of our current understanding of the genetic regulation of paternal care via olfaction in rodents comes from studies of laboratory mice, future studies should begin to explore what role, if any, these genes play in the regulation and expression of paternal care in naturally biparental species, such as prairie voles and Djungarian hamsters.

Funding: This research received no external funding. Thanks to James Cook University for administrative support.

Acknowledgments: Special thanks to David Wilson and Misha Rowell for proof-reading and providing constructive criticism. Thanks to an anonymous reviewer who provided constructive and insightful feedback on the manuscript.

Conflicts of Interest: The author declares no conflict of interest.

References

1. Nakahara, T.S.; Cardozo, L.M.; Ibarra-Soria, X.; Bard, A.D.; Carvalho, V.M.A.; Trintinalia, G.Z.; Logan, D.W.; Papes, F. Detection of pup odors by non-canonical adult vomeronasal neurons expressing an odorant receptor gene is influenced by sex and parenting status. *BMC Biol.* **2016**, *14*, 12. [CrossRef]
2. Brennan, P.A.; Keverne, E.B. Something in the air? New insights into mammalian pheromones. *Curr. Biol.* **2004**, *14*, R81–R89. [CrossRef]
3. Swaney, W.T.; Curley, J.P.; Champagne, F.A.; Keverne, E.B. The paternally expressed Gene *Peg3* regulates sexual experience-dependent preferences for estrous odors. *Behav. Neurosci.* **2008**, *122*, 963–973. [CrossRef]
4. Phillips, M.L.; Tang-Martinez, Z. Parent–offspring discrimination in the prairie vole and the effects of odors and diet. *Can. J. Zool.* **1998**, *76*, 711–716. [CrossRef]
5. Been, L.E.; Petrulis, A. Chemosensory and hormone information are relayed directly between the medial amygdala, posterior bed nucleus of the stria terminalis, and medial preoptic area in male Syrian hamsters. *Horm. Behav.* **2011**, *59*, 536–548. [CrossRef] [PubMed]
6. Rymer, T.L.; Pillay, N. An integrated understanding of paternal care in mammals: Lessons from the rodents. *J. Zool.* **2018**, *306*, 69–76. [CrossRef]
7. Schradin, C.; Anzenberger, G. Costs of infant carrying in common marmosets, *Callithrix jacchus*: An experimental analysis. *Anim. Behav.* **2001**, *62*, 289–295. [CrossRef]
8. Campbell, J.C.; Laugero, K.D.; Van Westerhuyzen, J.A.; Hostetler, C.M.; Cohen, J.D.; Bales, K.L. Costs of pair-bonding and paternal care in male prairie voles (*Microtus ochrogaster*). *Physiol. Behav.* **2009**, *98*, 367–373. [CrossRef] [PubMed]
9. Houston, A.I.; Székely, T.; McNamara, J.M. Conflict between parents over care. *Trends Ecol. Evol.* **2005**, *20*, 33–38. [CrossRef] [PubMed]
10. Getz, L.L.; McGuire, B. A comparison of living singly and in male–female pairs in the prairie vole, *Microtus ochrogaster*. *Ethology* **1993**, *94*, 265–278. [CrossRef]
11. Maynard Smith, J. Parental investment: A prospective analysis. *Anim. Behav.* **1977**, *25*, 1–9. [CrossRef]
12. Gubernick, D.J.; Teferi, T. Adaptive significance of male parental care in a monogamous mammal. *Proc. Roy. Soc. Lond. B* **2000**, *267*, 147–150. [CrossRef] [PubMed]
13. Tinbergen, N. On aims and methods of ethology. *Z. Tierpsychol.* **1963**, 410–433. [CrossRef]
14. Runcie, M.J. Biparental care and obligate monogamy in the rock-haunting possum, *Petropseudes dahli*, from tropical Australia. *Anim. Behav.* **2000**, *59*, 1001–1008. [CrossRef]
15. Lee, A.W.; Brown, R.E. Medial preoptic lesions disrupt parental behavior in both male and female California mice (*Peromyscus californicus*). *Behav. Neurosci.* **2002**, *116*, 968–975. [CrossRef]
16. Kohl, J.; Autry, A.E.; Dulac, C. The neurobiology of parenting: A neural circuit perspective. *Bioessays* **2016**, *39*, 1600159. [CrossRef]
17. Krettek, J.E.; Price, J.L. Amygdaloid projections to subcortical structures within the basal forebrain and brainstem in the rat and cat. *J. Comp. Neurol.* **1978**, *178*, 225–254. [CrossRef]
18. Simerly, R.B.; Swanson, L.W. The organization of neural inputs to the medial preoptic nucleus of the rat. *J. Comp. Neurol.* **1986**, *246*, 312–342. [CrossRef]
19. Numan, M. Neural basis of maternal behavior in the rat. *Psychoneuroendocrinology* **1988**, *13*, 47–62. [CrossRef]
20. Numan, M.; Corodimas, K.P.; Numan, M.J.; Factor, E.M.; Piers, W.D. Axon-sparing lesions of the preoptic region and substantia innominata disrupt maternal behavior in rats. *Behav. Neurosci.* **1988**, *102*, 381–396. [CrossRef]
21. Tsuneoka, Y.; Maruyama, T.; Yoshida, S.; Nishimori, K.; Kato, T.; Numan, M.; Kuroda, K.O. Functional, anatomical, and neurochemical differentiation of medial preoptic area subregions in relation to maternal behavior in the mouse. *J. Comp. Neurol.* **2013**, *521*, 1633–1663. [CrossRef] [PubMed]
22. Wu, Z.; Autry, A.E.; Bergan, J.F.; Watabe-Uchida, M.; Dulac, C.G. Galanin neurons in the medial preoptic area govern parental behaviour. *Nature* **2014**, *509*, 325–330. [CrossRef] [PubMed]
23. Kuroda, K.O.; Tachikawa, K.; Yoshida, S.; Tsuneoka, Y.; Numan, M. Neuromolecular basis of parental behavior in laboratory mice and rats: With special emphasis on technical issues of using mouse genetics. *Prog. Neuro-Psychopharmacol. Biol. Psych.* **2011**, *35*, 1205–1231. [CrossRef] [PubMed]
24. Swanson, L.W.; Petrovich, G.D. What is the amygdala? *Trends Neurosci.* **1998**, *21*, 323–331. [CrossRef]

25. Scalia, F.; Winans, S.S. The differential projections of the olfactory bulb and accessory olfactory bulb in mammals. *J. Comp. Neurol.* **1975**, *161*, 31–56. [CrossRef]
26. Bargmann, C.I. Comparative chemosensation from receptors to ecology. *Nature* **2006**, *444*, 295–301. [CrossRef]
27. Buck, L.; Axel, R. A novel multigene family may encode odorant receptors: A molecular basis for odor recognition. *Cell* **1991**, *65*, 175–187. [CrossRef]
28. Dulac, C.; Axel, R. A novel family of genes encoding putative pheromone receptors in mammals. *Cell* **1995**, *83*, 195–206. [CrossRef]
29. Zhang, X.; Rodriguez, I.; Mombaerts, P.; Firestein, S. Odorant and vomeronasal receptor genes in two mouse genome assemblies. *Genomics* **2004**, *83*, 802–811. [CrossRef]
30. Swaney, W.T.; Keverne, E.B. The evolution of pheromonal communication. *Behav. Brain Res.* **2009**, *200*, 239–247. [CrossRef]
31. Jiao, H.; Hong, W.; Nevo, E.; Li, K.; Zhao, H. Convergent reduction of *V1R* genes in subterranean rodents. *BMC Evol. Biol.* **2019**, *19*, 176. [CrossRef] [PubMed]
32. Brechbuhl, J.; Klaey, M.; Broillet, M.C. Grueneberg ganglion cells mediate alarm pheromone detection in mice. *Science* **2008**, *321*, 1092–1095. [CrossRef] [PubMed]
33. Breer, H.; Fleischer, J.; Strotmann, J. The sense of smell: Multiple olfactory subsystems. *Cell. Mol. Life Sci.* **2006**, *63*, 1465–1475. [CrossRef] [PubMed]
34. Luo, M.; Fee, M.S.; Katz, L.C. Encoding pheromonal signals in the accessory bulb of behaving mice. *Science* **2003**, *299*, 1196–1201. [CrossRef] [PubMed]
35. Rouquier, S.; Blancher, A.; Giorgi, D. The olfactory receptor gene repertoire in primates and mouse: Evidence for reduction of the functional fraction in primates. *Proc. Natl. Acad. Sci. USA* **2000**, *97*, 2870–2874. [CrossRef] [PubMed]
36. Nagai, M.H.; Armelin-Correa, L.M.; Malnic, B. Mongenic and monoallelic expression of odorant receptors. *Mol. Pharmacol.* **2016**, *90*, 633–639. [CrossRef] [PubMed]
37. Feinstein, P.; Bozza, T.; Rodriguez, I.; Vassalli, A.; Mombaerts, P. Axon guidance of mouse olfactory sensory neurons by odorant receptors and the ß2 adrenergic receptor. *Cell* **2004**, *117*, 833–846. [CrossRef]
38. Ressler, K.J.; Sullivan, S.L.; Buck, L.B. Information coding in the olfactory system: Evidence for a stereotyped and highly organized epitope map in the olfactory bulb. *Cell* **1994**, *79*, 1245–1255. [CrossRef]
39. Herrada, G.; Dulac, C. A novel family of putative pheromone receptors in mammals with a topographically organized and sexually dimorphic distribution. *Cell* **1997**, *90*, 763–773. [CrossRef]
40. Kambere, M.B.; Lane, R.P. Co-regulation of a large and rapidly evolving repertoire of odorant receptor genes. *BMC Neurosc.* **2007**, *8*, S2. [CrossRef]
41. Sato, T.; Hirono, J.; Hamana, H.; Ishikawa, T.; Shimizu, A.; Takashima, I.; Kajiwara, R.; Iijima, T. Architecture of odor information processing in the olfactory system. *Anat. Sci. Int.* **2008**, *83*, 195–206. [CrossRef] [PubMed]
42. Haberly, L.B. Parallel-distributed processing in olfactory cortex: New insights from morphological and physiological analysis of neuronal circuitry. *Chem. Senses* **2001**, *26*, 551–576. [CrossRef] [PubMed]
43. Kritzer, M. The distribution of immunoreactivity for intracellular androgen receptors in the cerebral cortex of hormonally intact adult male and female rats: Localization in pyramidal neurons making corticocortical connections. *Cereb. Cortex* **2004**, *14*, 268–280. [CrossRef] [PubMed]
44. Kirkpatrick, B.; Williams, J.R.; Slotnick, B.M.; Carter, C.S. Olfactory bulbectomy decreases social behavior in male prairie voles (*M. ochrogaster*). *Physiol. Behav.* **1994**, *55*, 885–889. [CrossRef]
45. Keverne, E. The vomeronasal organ. *Science* **1999**, *286*, 716–720. [CrossRef] [PubMed]
46. Li, C.-S., Kaba, H.; Saito, H.; Seto, K. Neural mechanisms underlying the action of primer pheromones in mice. *Neuroscience* **1990**, *36*, 773–778. [CrossRef]
47. Halpern, M. The organization and function of the vomeronasal system. *Annu. Rev. Neurosci.* **1987**, *10*, 325–362. [CrossRef]
48. Holy, T.E.; Dulac, C.; Meister, M. Responses of vomeronasal neurons to natural stimuli. *Science* **2000**, *289*, 1569–1572. [CrossRef]
49. Grus, W.E.; Shi, P.; Zhang, Y.-P.; Zhang, J. Dramatic variation of the vomeronasal pheromone gene repertoire among five orders of placental and marsupial mammals. *Proc. Natl. Acad. Sci. USA* **2005**, *102*, 5767–5772. [CrossRef]

50. Beny, Y.; Kimchi, T. Innate and learned aspects of pheromone-mediated social behaviours. *Anim. Behav.* **2014**, *97*, 301–311. [CrossRef]
51. Kevetter, G.A.; Winans, S.S. Connections of the corticomedial amygdala in the golden hamster. I. Efferents of the "vomeronasal amygdala". *J. Comp. Neurol.* **1981**, *197*, 81–98. [CrossRef] [PubMed]
52. Fernandez-Fewell, G.D.; Meredith, M. c-Fos expression in vomeronasal pathways of mated or pheromone-stimulated male golden hamsters: Contributions from vomeronasal sensory input and expression related to mating performance. *J. Neurosci.* **1994**, *14*, 3643–3654. [CrossRef] [PubMed]
53. Segovia, S.; Garcia-Falgueras, A.; Carrillo, B.; Collado, P.; Pinos, H.; Perez-Laso, C.; Vinader-Caerols, C.; Beyer, C.; Guillamon, A. Sexual dimorphism in the vomeronasal system of the rabbit. *Brain Res.* **2006**, *1102*, 52–62. [CrossRef] [PubMed]
54. Hines, M.; Allen, L.S.; Gorski, R.A. Sex differences in subregions of the medial nucleus of the amygdala and the bed nucleus of the stria terminalis of the rat. *Brain Res.* **1992**, *579*, 321–326. [CrossRef]
55. del Abril, A.; Segovia, S.; Guillamón, A. The bed nucleus of the stria terminalis in the rat: Regional sex differences controlled by gonadal steroids early after birth. *Dev. Brain Res.* **1987**, *32*, 295–300. [CrossRef]
56. Dulac, C.; O'Connell, L.A.; Wu, Z. Neural control of maternal and paternal behaviors. *Science* **2014**, *345*, 765–770. [CrossRef]
57. De Vries, G.J.; Villalba, C. Brain sexual dimorphism and sex differences in parental and other social behaviors. *Ann. NY Acad. Sci.* **1997**, *807*, 273–286. [CrossRef]
58. Tachikawa, K.S.; Yoshihara, Y.; Kuroda, K.O. Behavioral transition from attack to parenting in male mice: A crucial role of the vomeronasal system. *J. Neurosci.* **2013**, *33*, 5120–5126. [CrossRef]
59. Chess, A.; Simon, I.; Cedar, H.; Axel, R. Allelic inactivation regulates olfactory receptor gene expression. *Cell* **1994**, *78*, 823–834. [CrossRef]
60. Serizawa, S.; Miyamichi, K.; Nakatani, H.; Suzuki, M.; Saito, M.; Yoshihara, Y.; Sakano, H. Negative feedback regulation ensures the one receptor-one olfactory neuron rule in mouse. *Science* **2003**, *302*, 2088–2094. [CrossRef]
61. Malnic, B.; Hirono, J.; Sato, T.; Buck, L.B. Combinatorial receptor codes for odors. *Cell* **1999**, *96*, 713–723. [CrossRef]
62. Young, J.M.; Shykind, B.M.; Lane, R.P.; Tonnes-Priddy, L.; Ross, J.A.; Walker, M.; Williams, E.M.; Trask, B.J. Odorant receptor expressed sequence tags demonstrate olfactory expression of over 400 genes, extensive alternate splicing and unequal expression levels. *Genome Biol.* **2003**, *4*, R71. [CrossRef] [PubMed]
63. Bear, D.M.; Lassance, J.-M.; Hoekstra, H.E.; Datta, S.R. The evolving neural and genetic architecture of vertebrate olfaction. *Curr. Biol.* **2016**, *26*, R1039–R1049. [CrossRef] [PubMed]
64. Liberles, S.D.; Buck, L.B. A second class of chemosensory receptors in the olfactory epithelium. *Nature* **2006**, *442*, 645–650. [CrossRef] [PubMed]
65. Johnson, M.A.; Tsai, L.; Roy, D.S.; Valenzuela, D.H.; Mosley, C.; Magklara, A.; Lomvardas, S.; Liberles, S.D.; Barnea, G. Neurons expressing trace amine-associated receptors project to discrete glomeruli and constitute an olfactory subsystem. *Proc. Natl. Acad. Sci. USA* **2012**, *109*, 13410–13415. [CrossRef] [PubMed]
66. Liberles, S.D. Trace amine-associated receptors are olfactory receptors in vertebrates. *Ann. NY Acad. Sci.* **2009**, *1170*, 168–172. [CrossRef]
67. Hervé, D.; Le Moine, C.; Corvol, J.C.; Belluscio, L.; Ledent, C.; Fienberg, A.A.; Jaber, M.; Studler, J.M.; Girault, J.A. Galpha(olf) levels are regulated by receptor usage and control dopamine and adenosine action in the striatum. *J. Neurosci.* **2001**, *21*, 4390–4399. [CrossRef]
68. Belluscio, L.; Gold, G.H.; Nemes, A.; Axel, R. Mice deficient in G(olf) are anosmic. *Neuron* **1998**, *20*, 69–81. [CrossRef]
69. Greer, P.L.; Bear, D.M.; Lassance, J.-M.; Bloom, M.L.; Tsukahara, T.; Pashkovski, S.L.; Masuda, F.K.; Nowlan, A.C.; Kirchner, R.; Hoekstra, H.E.; et al. A family of non-GPCR chemosensors defines an alternative logic for mammalian olfaction. *Cell* **2016**, *165*, 1734–1748. [CrossRef]
70. Munger, S.D.; Leinders-Zufall, T.; McDougall, L.M.; Cockerham, R.E.; Schmid, A.; Wandernoth, P.; Wennemuth, G.; Biel, M.; Zufall, F.; Kelliher, K.R. An olfactory subsystem that detects carbon disulfide and mediates food-related social learning. *Curr. Biol.* **2010**, *20*, 1438–1444. [CrossRef]
71. Meredith, M. Chronic recording of vomeronasal pump activation in awake behaving hamsters. *Physiol. Behav.* **1994**, *56*, 345–354. [CrossRef]

72. Leinders-Zufall, T.; Lane, A.P.; Puche, A.C.; Ma, W.; Novotny, M.V.; Shipley, M.T.; Zufall, F. Ultrasensitive pheromone detection by mammalian vomeronasal neurons. *Nature* **2000**, *405*, 792–796. [CrossRef] [PubMed]
73. Grus, W.E.; Zhang, J. Distinct evolutionary patterns between chemoreceptors of 2 vertebrate olfactory systems and the differential tuning hypothesis. *Mol. Biol. Evol.* **2008**, *25*, 1593–1601. [CrossRef] [PubMed]
74. Ryba, N.J.P.; Tirindelli, R. A new multigene family of putative pheromone receptors. *Neuron* **1997**, *19*, 371–379. [CrossRef]
75. Mundy, N.I. Genetic basis of olfactory communication in primates. *Am. J. Primatol.* **2006**, *68*, 559–567. [CrossRef]
76. Halpern, M.; Jia, C.; Shapiro, L.S. Segregated pathways in the vomeronasal system. *Microsc. Res. Techniq.* **1998**, *41*, 519–529. [CrossRef]
77. von Campenhausen, H.; Mori, K. Convergence of segregated pheromonal pathways from the accessory olfactory bulb to the cortex in the mouse. *Eur. J. Neurosci.* **2000**, *12*, 33–46. [CrossRef]
78. Rodriguez, I.; Del Punta, K.; Rothman, A.; Ishii, T.; Mombaerts, P. Multiple new and isolated families within the mouse superfamily of V1r vomeronasal receptors. *Nature Neurosci.* **2002**, *5*, 134–140. [CrossRef]
79. Del Punta, K.; Leinders-Zufall, T.; Rodriguez, I.; Jukam, D.; Wysocki, C.J.; Ogawa, S.; Zufall, F.; Mombaerts, P. Deficient pheromone responses in mice lacking a cluster of vomeronasal receptor genes. *Nature* **2002**, *419*, 70–74. [CrossRef]
80. Loconto, J.; Papes, F.; Chang, E.; Stowers, L.; Jones, E.P.; Takada, T.; Kumánovics, A.; Lindahl, K.F.; Dulac, C. Functional expression of murine V2R pheromone receptors involves selective association with the M10 and M1 families of MHC class Ib molecules. *Cell* **2003**, *112*, 607–618. [CrossRef]
81. Dulac, C.; Torello, A.T. Molecular detection of pheromone signals in mammals: From genes to behaviour. *Nature Rev. Neurosci.* **2003**, *4*, 551–562. [CrossRef] [PubMed]
82. Ishii, T.; Hirota, J.; Mombaerts, P. Combinatorial coexpression of neural and immune multigene families in mouse vomeronasal sensory neurons. *Curr. Biol.* **2003**, *13*, 394–400. [CrossRef]
83. Rivière, S.; Challet, L.; Fluegge, D.; Spehr, M.; Rodriguez, I. Formyl peptide receptor-like proteins are a novel family of vomeronasal chemosensors. *Nature* **2009**, *459*, 574–577. [CrossRef]
84. Dietschi, Q.; Tuberosa, J.; Rösingh, L.; Loichot, G.; Ruedi, M.; Carleton, A.; Rodriguez, I. Evolution of immune chemoreceptors into sensors of the outside world. *Proc. Natl. Acad. Sci. USA* **2017**, *114*, 7397–7402. [CrossRef]
85. Dalton, R.P.; Lomvardas, S. Chemosensory receptor specificity and regulation. *Annu. Rev. Neurosci.* **2015**, *8*, 331–349. [CrossRef] [PubMed]
86. Champagne, F.A.; Curley, J.P. Genetics and epigenetics of parental care. In *The Evolution of Parental Care*; Royle, N.J., Smiseth, P.T., Kölliker, M., Eds.; Oxford University Press: Oxford, UK, 2012; pp. 304–324.
87. Schwagmeyer, P.L. Ground squirrel kin recognition abilities: Are there social and life-history correlates? *Behav. Genet.* **1988**, *18*, 495–510. [CrossRef] [PubMed]
88. Widdig, A. Paternal kin discrimination: The evidence and likely mechanisms. *Biol. Rev.* **2007**, *82*, 319–334. [CrossRef] [PubMed]
89. Todrank, J.; Heth, G.; Johnston, R.E. Kin recognition in golden hamsters: Evidence for kinship odour. *Anim. Behav.* **1998**, *55*, 377–386. [CrossRef]
90. Chamero, P.; Marton, T.F.; Logan, D.W.; Flanagan, K.; Cruz, J.R.; Saghatelian, A.; Cravatt, B.F.; Stowers, L. Identification of protein pheromones that promote aggressive behaviour. *Nature* **2007**, *450*, 899–902. [CrossRef]
91. Chamero, P.; Katsoulidou, V.; Hendrix, P.; Bute, R.; Roberts, R.; Matsunami, H.; Abramowitz, J.; Birnbaumer, L.; Zufall, F.; Leinders-Zufall, T. G protein Gαo is essential for vomeronasal function and aggressive behavior in mice. *Proc. Natl. Acad. Sci. USA* **2011**, *108*, 12898–12903. [CrossRef]
92. Kaur, A.W.; Ackels, T.; Kuo, T.-H.; Cichy, A.; Dey, S.; Hays, C.; Kateri, M.; Logan, D.W.; Marton, T.F.; Spehr, M.; et al. Murine pheromone proteins constitute a context-dependent combinatorial code governing multiple social behaviors. *Cell* **2014**, *157*, 676–688. [CrossRef] [PubMed]
93. Trouillet, A.-C.; Keller, M.; Weiss, J.; Leinders-Zufall, T.; Birnbaumer, L.; Zufall, F.; Chamero, P. Central role of G protein Gαi2 and Gαi2$^+$ vomeronasal neurons in balancing territorial and infant-directed aggression of male mice. *Proc. Natl. Acad. Sci. USA* **2019**, *116*, 5135–5143. [CrossRef] [PubMed]
94. Lévai, O.; Feistel, T.; Breer, H.; Strotman, J. Cells in the vomeronasal organ express odorant receptors but project to the accessory olfactory bulb. *J. Comp. Neurol.* **2006**, *498*, 476–490. [CrossRef]

95. Leypold, B.G.; Yu, C.R.; Leinders-Zufall, T.; Kim, M.M.; Zufall, F.; Axel, R. Altered sexual and social behaviors in trp2 mutant mice. *Proc. Natl. Acad. Sci. USA* **2002**, *99*, 6376–6381. [CrossRef]
96. Stowers, L.; Holy, T.E.; Meister, M.; Dulac, C.; Koentges, G. Loss of sex discrimination and male-male aggression in mice deficient for TRP2. *Science* **2002**, *295*, 1493–1500. [CrossRef] [PubMed]
97. Spehr, M.; Kelliher, K.R.; Li, X.-H.; Boehm, T.; Leinders-Zufall, T.; Zufall, F. Essential role of the main olfactory system in social recognition of major histocompatibility complex peptide ligands. *J. Neurosci.* **2006**, *26*, 1961–1970. [CrossRef]
98. Hasen, N.S.; Gammie, S.C. *Trpc2* gene impacts on maternal aggression, accessory olfactory bulb anatomy and brain activity. *Genes Brain Behav.* **2009**, *8*, 639–649. [CrossRef]
99. Kimchi, T.; Xu, J.; Dulac, C. A functional circuit underlying male sexual behvior in the female mouse brain. *Nature* **2007**, *448*, 1009–1014. [CrossRef]
100. Jin, D.; Liu, H.-X.; Hirai, H.; Torashima, T.; Nagai, T.; Lopatina, O.; Shnayder, N.A.; Yamada, K.; Noda, M.; Seike, T.; et al. CD38 is critical for social behaviour by regulating oxytocin secretion. *Nature* **2007**, *446*, 41–45. [CrossRef]
101. Akther, S.; Korshnova, N.; Zhong, J.; Liang, M.; Cherepanov, S.M.; Lopatina, O.; Komleva, Y.K.; Salmina, A.B.; Nishimura, T.; Fakhrul, A.A.K.M.; et al. CD38 in the nucleus accumbens and oxytocin are related to paternal behavior in mice. *Mol. Brain* **2013**, *6*, 41. [CrossRef]
102. Grigor'eva, M.E.; Golubeva, M.G. Oxytocin: Structure, synthesis, receptors, and basic effects. *Neurochem. J.* **2010**, *4*, 75–83. [CrossRef]
103. Marsh, A.A.; Yu, H.H.; Pine, D.S.; Gorodetsky, E.K.; Goldman, D.; Blair, R.J.R. The influence of oxytocin administration on responses to infant faces and potential moderation by *OXTR* genotype. *Psychopharmacology* **2012**, *224*, 469–476. [CrossRef] [PubMed]
104. Barberis, C.; Tribollet, E. Vasopressin and oxytocin receptors in the central nervous system. *Crit. Rev. Neurobiol.* **1996**, *10*, 119–154. [CrossRef] [PubMed]
105. Cho, M.M.; DeVries, A.C.; Williams, J.R.; Carter, C.S. The effects of oxytocin and vasopressin on partner preferences in male and female prairie voles (*Microtus ochrogaster*). *Behav. Neurosci.* **1999**, *113*, 1071–1079. [CrossRef]
106. Parker, K.J.; Kinney, L.F.; Phillips, K.M.; Lee, T.M. Paternal behavior is associated with central neurohormone receptor binding patterns in meadow voles (*Microtus pennsylvanicus*). *Behav. Neurosci.* **2001**, *115*, 1341–1348. [CrossRef]
107. Lopatina, O.; Inzhutova, A.; Pichugina, Y.A.; Okamato, H.; Salmina, A.B.; Higashida, H. Reproductive experience affects parental retrieval behaviour associated with increased plasma oxytocin levels in wild-type and *CD38*-knockout mice. *J. Neuroendocrinol.* **2011**, *23*, 1125–1133. [CrossRef]
108. Song, Z.; Tai, F.; Yu, C.; Wu, R.; Zhang, X.; Broders, H.; He, F.; Guo, R. Sexual or paternal experiences alter alloparental behavior and the central expression of ERα and OT in male mandarin voles (*Microtus mandarinus*). *Behav. Brain Res.* **2010**, *214*, 290–300. [CrossRef]
109. Lambert, K.G.; Franssen, C.L.; Hampton, J.E.; Rzucidlo, A.M.; Hyer, M.M.; True, M.; Kaufman, C.; Bardi, M. Modeling paternal attentiveness: Distressed pups evoke differential neurobiological and behavioral responses in paternal and nonpaternal mice. *Neuroscience* **2013**, *234*, 1–12. [CrossRef]
110. Horrell, N.D.; Hickmott, P.W.; Saltzman, W. Neural regulation of paternal behaviour in mammals: Sensory, neuroendocrine, and experiential influences on the paternal brain. *Curr. Top. Behav. Neurosci.* **2019**, *43*, 111–160. [CrossRef]
111. Svare, B.; Mann, M. Infanticide: Genetic, developmental and hormonal influences in mice. *Physiol. Behav.* **1981**, *27*, 921–927. [CrossRef]
112. Tsuneoka, Y.; Tokita, K.; Yoshihara, C.; Amano, T.; Esposito, G.; Huang, A.J.; Yu, L.M.; Odaka, Y.; Shinozuka, K.; McHugh, T.J.; et al. Distinct preoptic-BST nuclei dissociate paternal and infanticidal behavior in mice. *EMBO J.* **2015**, *34*, 2652–2670. [CrossRef] [PubMed]
113. Böcskei, Z.; Groom, C.R.; Flower, D.R.; Wright, C.E.; Phillips, S.E.V.; Cavaggioni, A.; Findlay, J.B.C.; North, A.C.T. Pheromone binding to two rodent urinary proteins revealed by X-ray crystallography. *Nature* **1992**, *360*, 186–188. [CrossRef] [PubMed]
114. Beynon, R.J.; Hurst, J.L. Multiple roles of major urinary proteins in the house mouse, *Mus domesticus*. *Biochem. Soc. Trans.* **2003**, *31*, 142–146. [CrossRef] [PubMed]
115. Flower, D.R. The lipocalin protein family: Structure and function. *Biochem. J.* **1996**, *318*, 1–14. [CrossRef]

116. Al-Shawi, R.; Ghazal, P.; Clark, A.J.; Bishop, J.O. Intraspecific evolution of a gene family coding for urinary proteins. *J. Mol. Evol.* **1989**, *29*, 302–313. [CrossRef] [PubMed]
117. Utsumi, M.; Ohno, K.; Kawasaki, Y.; Tamura, M.; Kubo, T.; Tohyama, M. 1999. Expression of major urinary protein genes in the nasal glands associated with general olfaction. *J. Neurobiol.* **1999**, *39*, 227–236. [CrossRef]
118. Cavaggioni, A.; Mucignat-Caretta, C. Major urinary proteins, α_{2U}-globulins and aphrodisin. *Biochim. Biophys. Acta* **2000**, *1482*, 218–228. [CrossRef]
119. Busquet, N.; Baudoin, C. Odour similarities as a basis for discriminating degrees of kinship in rodents: Evidence from *Mus spicilegus*. *Anim. Behav.* **2005**, *70*, 997–1002. [CrossRef]
120. Jenkins, R.; Tetzlaff, W.; Hunt, S.P. Differential expression of immediate early genes in rubrospinal neurons following axotomy in rat. *Eur. J. Neurosci.* **1993**, *5*, 203–209. [CrossRef]
121. Moffatt, C.A.; Ball, G.F.; Nelson, R.J. The effects of photoperiod on olfactory c-*fos* expression in prairie voles, *Microtus ochrogaster*. *Brain Res.* **1995**, *677*, 82–88. [CrossRef]
122. Fiber, J.M.; Adames, P.; Swann, J.M. Pheromones induce c-*fos* in limbic areas regulating male hamster mating behaviour. *NeuroReport* **1993**, *4*, 871–874. [CrossRef] [PubMed]
123. Brown, J.R.; Ye, H.; Bronson, R.T.; Dikkes, P.; Greenberg, M.E. A defect in nurturing in mice lacking the immediate early gene *fosB*. *Cell* **1996**, *86*, 297–309. [CrossRef]
124. Kuroda, K.O.; Meaney, M.J.; Uetani, N.; Fortin, Y.; Ponton, A.; Kato, T. ERK-FosB signaling in dorsal MPOA neurons plays a major role in the initiation of parental behavior in mice. *Mol. Cell. Neurosci.* **2007**, *36*, 121–131. [CrossRef] [PubMed]
125. Stolzenberg, D.S.; Mayer, H.S. Experience-dependent mechanisms in the regulation of parental care. *Front. Neuroendocrinol.* **2019**, *54*, 100745. [CrossRef]
126. Jin, S.-H.; Blendy, J.A.; Thomas, S.A. Cyclic AMP response element-binding protein is required for normal maternal nurturing behavior. *Neuroscience* **2005**, *133*, 647–655. [CrossRef]
127. Jones, D.T.; Reed, R.R. G_{olf}: An olfactory neuron specific-G protein involved in odorant signal transduction. *Science* **1989**, *244*, 790–795. [CrossRef]
128. Bakalyar, H.A.; Reed, R.R. Identification of a specialized adenylyl cyclase that may mediate odorant detection. *Science* **1990**, *250*, 1403–1406. [CrossRef]
129. Nakamura, T.; Gold, G.H. A cyclic nucleotide-gated conductance in olfactory receptor cilia. *Nature* **1987**, *325*, 442–444. [CrossRef]
130. Wang, Z.; Storm, D.R. Maternal behavior is impaired in female mice lacking type 3 adenylyl cyclase. *Neuropsychopharmacology* **2011**, *36*, 772–781. [CrossRef]
131. Kohl, J.; Babayan, B.M.; Rubinstein, N.D.; Autry, A.E.; Marin-Rodriguez, B.; Kapoor, V.; Miyamishi, K.; Zweifel, L.S.; Luo, L.; Uchida, N.; et al. Functional circuit architecture underlying parental behaviour. *Nature* **2018**, *556*, 326–331. [CrossRef]
132. Fang, Y.-Y.; Yamaguchi, T.; Song, S.C.; Tritsch, N.X.; Lin, D. A hypothalamic midbrain pathway essential for driving maternal behaviors. *Neuron* **2018**, *98*, 192–207. [CrossRef] [PubMed]
133. Kuroda, K.O.; Tsuneoka, Y. Assessing postpartum maternal care, alloparental behavior, and infanticide in mice: With notes on chemosensory influences. *Methods Mol. Biol.* **2013**, *1068*, 331–347. [CrossRef] [PubMed]
134. Wong, S.T.; Trinh, K.; Hacker, B.; Chan, G.C.; Lowe, G.; Gaggar, A.; Xia, Z.; Gold, G.H.; Storm, D.R. Disruption of the type III adenylyl cyclase gene leads to peripheral and behavioral anosmia in transgenic mice. *Neuron* **2000**, *27*, 487–497. [CrossRef]
135. Hervé, D. Identification of a specific assembly of the G protein Golf as a critical and regulated module of dopamine and adenosine-activated cAMP pathways in the striatum. *Front. Neuroanat.* **2011**, *5*, 1–9. [CrossRef] [PubMed]
136. Saltzman, W.; Maestripieri, D. The neuroendocrinology of primate maternal behavior. *Prog. Neuro-Psychopharmacol. Biol. Psych.* **2011**, *35*, 1192–1204. [CrossRef] [PubMed]
137. Grattan, D.R.; Kokay, I.C. Prolactin: A pleiotropic neuroendocrine hormone. *J. Neuroendocrinol.* **2008**, *20*, 752–763. [CrossRef]
138. Walsh, R.J.; Slaby, F.J.; Posner, B.I. A receptor-mediated mechanism for the transport of prolactin from blood to cerebrospinal fluid. *Endocrinology* **1987**, *120*, 1846–1850. [CrossRef]
139. Brown, R.E.; Murdoch, T.; Murphy, P.R.; Moger, W.H. Hormonal responses of male gerbils to stimuli from their mate and pups. *Horm. Behav.* **1995**, *29*, 474–491. [CrossRef]

140. Schradin, C.; Anzenberger, G. Prolactin, the hormone of paternity. *News Physiol. Sci.* **1999**, *14*, 223–231. [CrossRef]
141. Mak, G.K.; Weiss, S. Paternal recognition of adult offspring mediated by newly generated CNS neurons. *Nature Neurosci.* **2010**, *13*, 753–758. [CrossRef]
142. Ma, E.; Lau, J.; Grattan, D.R.; Lovejoy, D.A.; Wynne-Edwards, K.E. Male and female prolactin receptor mRNA expression in the brain of a biparental and a uniparental hamster, *Phodopus*, before and after the birth of a litter. *J. Neuroendocrinol.* **2005**, *17*, 81–90. [CrossRef] [PubMed]

MDPI
St. Alban-Anlage 66
4052 Basel
Switzerland
Tel. +41 61 683 77 34
Fax +41 61 302 89 18
www.mdpi.com

Genes Editorial Office
E-mail: genes@mdpi.com
www.mdpi.com/journal/genes

www.ingramcontent.com/pod-product-compliance
Lightning Source LLC
LaVergne TN
LVHW072009150826
845671LV00011B/306
* 9 7 8 3 0 3 9 3 6 6 2 1 7 *